P9-DTY-548

THE COMPETITIVE EDGE

Research Priorities for U.S. Manufacturing

Committee on Analysis of Research
Directions and Needs in U.S. Manufacturing

Manufacturing Studies Board
Commission on Engineering and Technical Systems
National Research Council

NATIONAL ACADEMY PRESS
Washington, D.C. 1991

NATIONAL ACADEMY PRESS • 2101 Constitution Avenue, N.W. • Washington, D.C. 20418

NOTICE: The project that is the subject of this report was approved by the Governing Board of the National Research Council, whose members are drawn from the councils of the National Academy of Sciences, the National Academy of Engineering, and the Institute of Medicine. The members of the committee responsible for the report were chosen for their special competences and with regard for appropriate balance.

This report has been reviewed by a group other than the authors according to procedures approved by a Report Review Committee consisting of members of the National Academy of Sciences, the National Academy of Engineering, and the Institute of Medicine.

This study was supported by Contract No. DMC-8717382 between the National Science Foundation and the National Academy of Sciences.

Library of Congress Cataloging-in-Publication Data

The Competitive edge : research priorities for U.S. manufacturing : report of the Committee on Analysis of Research Directions and Needs in U.S. Manufacturing, Manufacturing Studies Board, Commission on Engineering and Technical Systems, National Research Council.
p. cm.
Includes bibliographical references (p.) and index.
ISBN 0-309-04385-9 : $24.95
1. Production engineering—Research—United States. I. National Research Council (U.S.). Committee on Analysis of Research Directions and Needs in U.S. Manufacturing.
TS176.C6 1991 91-17465
658.5'072073—dc20 CIP

Printed in the United States of America

COMMITTEE ON ANALYSIS OF RESEARCH DIRECTIONS AND NEEDS IN U.S. MANUFACTURING

CYRIL M. PIERCE, *Chairman*, General Manager, Manufacturing and Quality Technology Department, GE Aircraft Engines, Cincinnati, Ohio
AVAK AVAKIAN, Vice President (retired), GTE Government Systems, Concord, Massachusetts
GERARDO BENI, Director, Center for Robotics Systems in Microelectronics, University of California, Santa Barbara
WILLIAM G. HOWARD, JR., Senior Fellow, National Academy of Engineering, Scottsdale, Arizona
RAMCHANDRAN JAIKUMAR, Professor of Business Administration, Graduate School of Business Administration, Harvard University, Boston, Massachusetts
JOEL MOSES, Dean of Engineering, Massachusetts Institute of Technology, Cambridge
GUSTAV J. OLLING, Chief, Automotive Research and CAD/CAM User Systems, Chrysler Corporation, Highland Park, Michigan
HRIDAY R. PRASAD, Manager of Technology Planning, North American Automotive Manufacturing Operations, Ford Motor Company, Dearborn, Michigan
A. TIM SHERROD, President, Savant Solutions Company, Menlo Park, California
DAN L. SHUNK, Director, CIM Systems Research Center, Arizona State University, Tempe
JAMES C. WILLIAMS, General Manager, Engineering Materials Technology Laboratory, General Electric Company, Cincinnati, Ohio
MICHAEL J. WOZNY, Director, Rensselaer Design Research Center, Rensselaer Polytechnic Institute, Troy, New York

Staff

VERNA J. BOWEN, Staff Assistant (from Dec. 17, 1989)
LUCY V. FUSCO, Staff Assistant
THOMAS C. MAHONEY, Acting Director
KAREN L. MILLAN, Staff Assistant (until Dec. 16, 1989)
KERSTIN B. POLLACK, Study Director; MSB Director for Program Development; MSB Deputy Director
JOHN SIMON, Consultant, Writer-Editor
ERIC A. THACKER, Research Associate (until Aug. 10, 1990)

MANUFACTURING STUDIES BOARD

JAMES F. LARDNER, *Chairman*, Vice President (Retired), Component Group, Deere & Company, Davenport, Iowa

MATTHEW O. DIGGS, JR., Chairman, The Diggs Group, Dayton, Ohio

CHARLES P. FLETCHER, Vice President of Engineering, Aluminum Company of America, Pittsburgh, Pennsylvania

HEINZ K. FRIDRICH, Vice President, Manufacturing, IBM Corporation, Purchase, New York

DAVID A. GARVIN, Professor, Business Administration, Graduate School of Business Administration, Harvard University, Boston, Massachusetts

LEONARD A. HARVEY, Secretary of Commerce, Labor, and Environmental Resources, State of West Virginia (Retired), Vienna, West Virginia

CHARLES W. HOOVER, JR., Professor, Department of Industrial and Mechanical Engineering, Polytechnic University, Brooklyn, New York

JOEL MOSES, Dean of Engineering, Massachusetts Institute of Technology, Cambridge

LAURENCE C. SEIFERT, Vice President, Communications and Computer Products, Sourcing and Manufacturing, AT&T Company, Bridgewater, New Jersey

JOHN M. STEWART, Director, McKinsey and Company, Inc., New York, New York

WILLIAM J. USERY, JR., President, Bill Usery Associates, Inc., Washington, D.C.

HERBERT B. VOELCKER, Charles Lake Professor of Engineering, Sibley School of Mechanical Engineering, Cornell University, Ithaca, New York

Staff

VERNA J. BOWEN, Staff Assistant

DANA G. CAINES, Staff Associate

LUCY V. FUSCO, Staff Assistant

THEODORE W. JONES, Research Associate

THOMAS C. MAHONEY, Acting Director

KERSTIN B. POLLACK, Deputy Director, and Director of New Program Development

MICHAEL L. WITMORE, Research Assistant

Preface

The Committee on Analysis of Research Directions and Needs in U.S. Manufacturing was charged by the National Science Foundation (NSF) with (1) identifying and ranking manufacturing-related technologies and disciplines to produce a comprehensive national manufacturing research agenda, and (2) performing in-depth analyses of some of the technologies and disciplines identified in that agenda. Concluding that the most important purpose of university efforts in manufacturing engineering and technology is to attract the most capable students to manufacturing careers, the committee determined that a national research agenda should address topics that encourage and develop students and faculty, while meeting industry's needs for new technology and high-leverage technical concepts. The committee reasoned that the audience for such an agenda extended beyond the program directors in the NSF's engineering directorate to the research community in government, industry, and academe.

The nature of manufacturing suggests a research spectrum ranging from concept definition to proof of concept feasibility to development of applications and implementation mechanisms. The committee reasoned that if problems are carefully selected and thoughtfully researched the results will provide the basis for practical application, and therefore agreed to focus on only the front end of this spectrum—definition and proof of concept. The committee nevertheless recognizes that improving the pipeline from concept to commercial viability is also an important issue in U.S. manufacturing.

The committee established four criteria for qualifying and ranking research:

1. each project should be researchable within NSF's or other government agencies' accepted guidelines, and results should be available within a reasonable time frame;
2. the research results should be useful in multiple industrial applications and provide capabilities and experience that advance manufacturing operations and competitiveness;
3. research results should promote fundamental change in management practice and culture; and
4. each project should expand scientific research relevant to manufacturing problems, encourage academic researchers to emphasize an interdisciplinary approach, and promote greater rapport between researchers and practitioners.

Interdisciplinary topics and problems in soft areas—e.g., management, human resources, and education—were deemed as important as the technology issues. The committee decided that the best way to handle soft issues is to encourage researchers to study new technologies and disciplines and their implications for managers, workers, and organizations concurrently. When shortcomings are evident in existing practices in these areas, however, the committee agreed to include relevant research topics in the comprehensive agenda.

The committee then subdivided manufacturing into six categories and nominated panels to develop research recommendations in each:

- intelligent manufacturing control,
- equipment reliability and maintenance,
- manufacturing of and with advanced engineered materials,
- manufacturing skills improvement,
- rapid product realization, and
- alternative concepts in manufacturing.

Subsequently, it was decided to narrow the categories further. Materials developed by the panel on alternative concepts in manufacturing were used to help develop an overview of the report and to add management and organizational issues to what were initially largely technology-oriented materials on the product realization process ("rapid" was dropped from the descriptor for that category because it was thought to be implicit). The final report thus recommends research in five general areas:

1. Intelligent Manufacturing Control
2. Equipment Reliability and Maintenance
3. Manufacturing of and with Advanced Engineered Materials
4. The Product Realization Process
5. Manufacturing Skills Improvement

CYRIL M. PIERCE, *Chairman*
Committee on Analysis of Research Directions and Needs in U.S. Manufacturing

The National Academy of Sciences is a private, nonprofit, self-perpetuating society of distinguished scholars engaged in scientific and engineering research, dedicated to the furtherance of science and technology and to their use for the general welfare. Upon the authority of the charter granted to it by the Congress in 1863, the Academy has a mandate that requires it to advise the federal government on scientific and technical matters. Dr. Frank Press is president of the National Academy of Sciences.

The National Academy of Engineering was established in 1964, under the charter of the National Academy of Sciences, as a parallel organization of outstanding engineers. It is autonomous in its administration and in the selection of its members, sharing with the National Academy of Sciences the responsibility for advising the federal government. The National Academy of Engineering also sponsors engineering programs aimed at meeting national needs, encourages education and research, and recognizes the superior achievements of engineers. Dr. Robert M. White is president of the National Academy of Engineering.

The Institute of Medicine was established in 1970 by the National Academy of Sciences to secure the services of eminent members of appropriate professions in the examination of policy matters pertaining to the health of the public. The Institute acts under the responsibility given to the National Academy of Sciences by its congressional charter to be an adviser to the federal government and, upon its own initiative, to identify issues of medical care, research, and education. Dr. Samuel O. Thier is president of the Institute of Medicine.

The National Research Council was organized by the National Academy of Sciences in 1916 to associate the broad community of science and technology with the Academy's purposes of furthering knowledge and advising the federal government. Functioning in accordance with general policies determined by the Academy, the Council has become the principal operating agency of both the National Academy of Sciences and the National Academy of Engineering in providing services to the government, the public, and the scientific and engineering communities. The Council is administered jointly by both Academies and the Institute of Medicine. Dr. Frank Press and Dr. Robert M. White are chairman and vice chairman, respectively, of the National Research Council.

Contents

THE COMPETITIVE EDGE

Executive Summary

ADVANCED MANUFACTURING technology is emerging as a major corporate advantage in world trade. Strategic application of such technology can markedly improve manufacturers' product quality, reponsiveness to customers, process control and flexibility, and flexibility of capital investment—all determinants of global manufacturing competitiveness.

Progress in U.S. manufacturing technologies and competitiveness faces significant barriers: inflexible organizations; inadequate technology; inappropriate performance measures; and lack of appreciation for the importance of manufacturing. These barriers are addressed in this report of the Committee on Analysis of Research Directions and Needs in U.S. Manufacturing, Manufacturing Studies Board, Commission on Engineering and Technical Systems, National Research Council. The report identifies and analyzes research needs in five critical areas of manufacturing: intelligent manufacturing control, equipment reliability and maintenance, advanced engineered materials, manufacturing skills improvement, and the product realization process.

Intelligent manufacturing control requires research in several areas. They include: sensor technology in data integration and pattern recognition; adaptable knowledge bases of design, manufacturing, and management intelligence; and creation of a dynamic model of manufacturing.

Equipment reliability and maintenance programs are underutilized in this country largely because of manufacturing managers' lack of awareness of their economic benefits. Technology problems associated with these programs are more tractable than the associated people problems.

Needs in advanced engineered materials include integration of processing methods into the design and development of new materials from the beginning. A need exists also to instill sensitivity to materials' properties into process design schemes.

The product realization process—from initial idea through marketing—has research needs in three areas: developing intelligent product images, establishing the requisite connections among them, and devising an organizational structure in which these concepts can be made operational.

Manufacturing skills improvement is critical to advanced manufacturing technology with its need for work-force skills that U.S. schools are neither cultivating nor preparing students to acquire. The first need is basic literacy. The goal is a manufacturing work force with multidisciplinary skills of a high order.

Fruitful pursuit of the recommended research could transform U.S. manufacturing. Potential results include:

- Highly specialized processing of metals, ceramics, and polymers to yield radical improvements in materials' strength and toughness;
- Equipment operators working synergistically with intelligent control systems that are capable of predicting, preventing, or automatically remedying equipment failure;
- Autonomous manufacturing control systems that exploit human powers of perception, pattern recognition, and problem-solving in conjunction with machine capacity for manipulating vast amounts of data;
- Global information systems that enable electronic virtual enterprises to access and coordinate the localized design and manufacturing capabilities of village industries all over the world; and
- Production handled by highly skilled professionals working as components of human–machine systems that are linked integrally with management functions.

To achieve such results the manufacturing community must learn from and adopt the fruits of the research proposed, and the nation must reinvigorate its educational system. Manufacturing would then come to be viewed as a national asset and careers in manufacturing would be highly regarded.

1

Overview

ADVANCED MANUFACTURING technology, because its strategic application results in important competitive advantages, including higher quality, greater responsiveness to consumer demands, and greater flexibility of capital investment, is emerging as a major corporate advantage in world trade. Development and deployment of advanced manufacturing technology will be affected by a host of factors, including capital, markets, tax and trade policy, corporate management, and skills.

In a world that has come to expect zero defects and high reliability, manufacturing[1] will move from a craft to a science, thereby requiring a much smaller work force that possesses multidisciplinary skills of a high order. Figures 1-1 to 1-3 show this projected decline in manufacturing employment to the year 2000. In Figure 1-1, manufacturing employment is shown to decline by almost 20 percent between 1950 and 2000. In Figure 1-2, employment in more highly skilled occupations (i.e., computer specialists, electrical engineers, etc.) shows an increase from 9 to 52 percent between 1988 and 2000, while less highly skilled job categories (i.e., machine tool operators) showed a 5- to 44-percent decline. And in Figure 1-3, manufacturing is shown to be the only occupational group to show declining employment, in contrast to groups such as professional and technical workers, which shows a 24-percent projected increase between 1988 and 2000. The new skills that this smaller work force will need are dictated by three characteristics of advanced manufacturing technology:

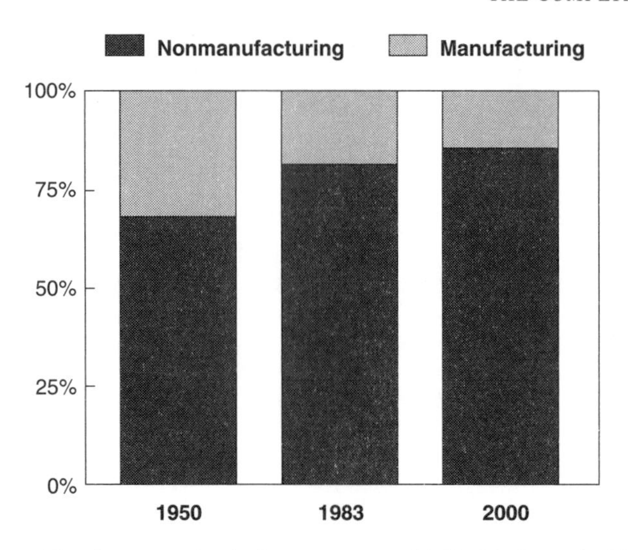

FIGURE 1-1 Employment by industry sector, 1950, 1983, and 2000. Source: W. B. Johnston and A.H. Packer. 1987. Workforce 2000: Work and workers for the 21st century. Hudson Institute. 51-73.

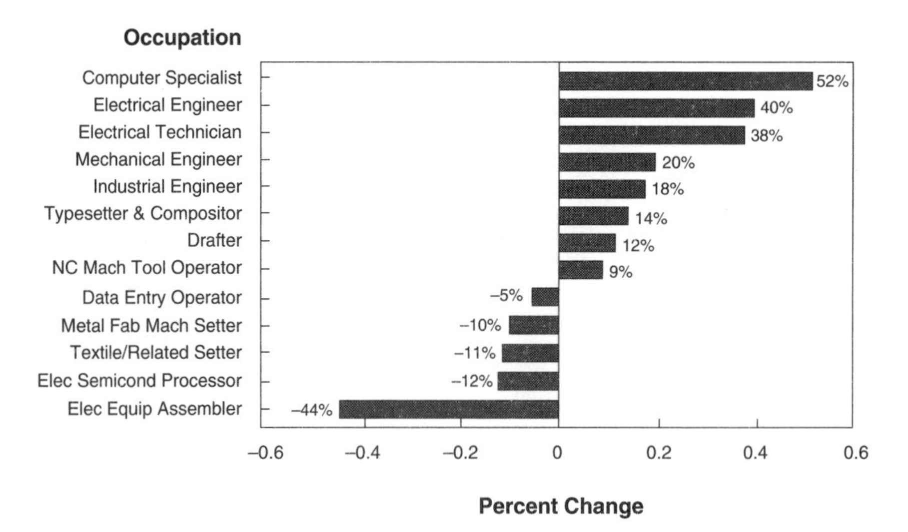

FIGURE 1-2 Projected employment changes, 1988-2000. Source: Projections of occupational employment, 1988-2000. 1989. G. Silvestri and J. Lukasiewicz. Monthly Labor Review (112:11): 42-65.

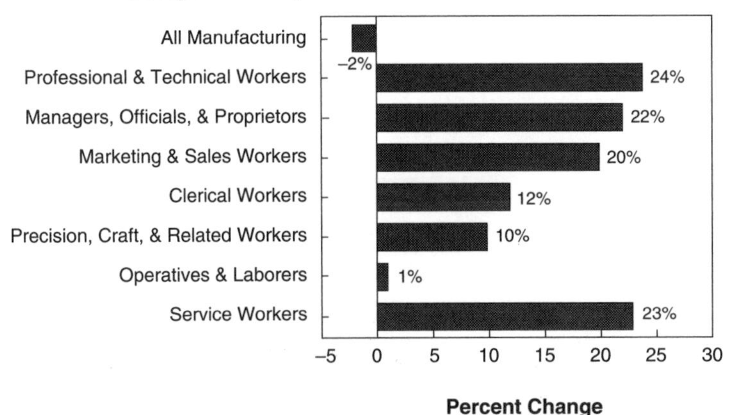

FIGURE 1-3 Projected manufacturing employment changes, 1988-2000. Source: Projections of occupational employment, 1988-2000. Source: G. Silvestri and J. Lukasiewicz. 1989. Monthly Labor Review (112:11):42-65.

- its integration of information and materials handling and processing, which are separate in traditional automation;
- its reliance on higher levels of human and machine intelligence rather than on the skill of operators; and
- its placement of the production process largely under the control of computer programs into which product and process specifications are fed as computer instructions.

Firms that use advanced manufacturing technology will thus be distinguished less by their manufacturing processes than by the integrated systems that drive those processes. Creative use of the technology will yield greater competitive advantage. That is, the benefits of advanced manufacturing technology will depend more heavily on the people who develop the systems that determine the range and types of products that can be produced and resolve the contingencies that arise in production, than on the hardware that is likely to be available to any company that can pay for it. Figure 1-4 shows for the period 1973 to 1983 the educational attainment in manufacturing as a percentage of all manufacturing employees.

The nature of the new skills required is perhaps best understood in terms of the changes that will accompany the development and deployment of advanced manufacturing technologies. Tight integration of information and materials processing (see Chapter

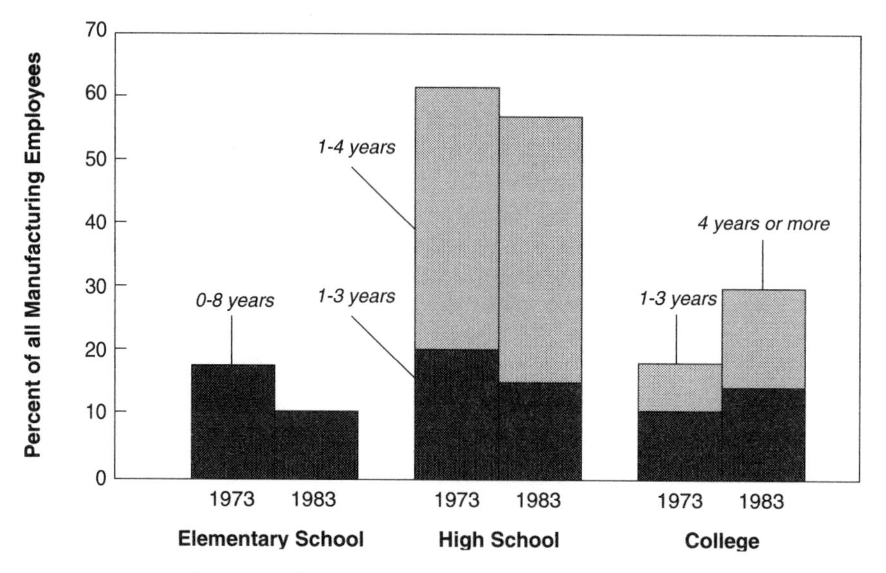

FIGURE 1-4 Educational attainment in manufacturing, 1973 and 1983. Source: Projections of occupational employment, 1988-2000. Source: G. Silvestri and J. Lukasiewicz. 1989. Monthly Labor Review (112:11):42-65.

5) will extend from choice of product, to design, to materials transformation and/or product assembly, and, finally, to marketing and distribution. Scheduling and resource allocation will be integrally linked to, and occur simultaneously with, planning in both the extremely short term (e.g., product time to market) and very long term (e.g., technology development and/or acquisition). Intelligent manufacturing control (see Chapter 2) will provide a means of controlling and recording continuous information about what is happening on the factory floor, and means of reporting the information on which operators and managers rely to make decisions. Control will be exerted primarily at the level of the work cell, will require significant intercell coordination, and will involve skills that have both human and machine components. Decision making in this context will require an understanding of logical relationships and statistical correlation and involve teamwork among peers. A growing portfolio of products and processes, machines and operators, and external forces will provide multiple windows for viewing problems, creating a concomitant demand for broader experience.

Time-to-market pressures that attend advanced manufacturing technology will drive organizations to compress their organizational hierarchy—become flatter—in order to push more responsibility down to the factory floor. Operators, for example, will be called on to perform tasks that formerly were divided among several workers with different job classifications (e.g., machine setup and adjustment and quality monitoring, as well as operation). To maintain production while performing the frequent adjustments that are required by complex, integrated manufacturing systems, operators will have to be able to analyze problems and implement solutions, often in the form of computer programming. Operators will increasingly control production rather than be controlled by it.

The growing importance of manufacturing skills is best seen in light of the significance of advanced manufacturing technology. Product and process specifications that exist as computer instructions are highly transportable. Such transportability, together with the flexibility inherent in advanced manufacturing technology, creates a context in which a product can be produced anywhere, at home or abroad, by anyone with the requisite hardware. This situation renders production capacity a commodity.

The Japanese metalworking industry, using advanced manufacturing technology, has demonstrated remarkable increases in productivity: fivefold reductions in labor, reductions by half in the number of machines required, increases in machine utilization of more than 20 percent, delivery performances of 100 percent over three months, unscheduled system downtimes of 2 percent, and quality problems at the level of 0.006 percent (see Tables 1-1 and 1-2). Advanced manufacturing technology makes traditional

TABLE 1-1 Comparison of Flexible Manufacturing Systems Studied in the United States and Japan

Manufacturing Systems Variables	United States	Japan
System development time (years)	2.5 to 3	1.25 to 1.75
Number of machines per system	7	6
Types of parts produced per system	10	93
Annual volume per part	1,727	258
Number of parts produced per day	88	120
Number of new parts introduced per year	1	22
Number of systems with untended operations	0	18
Utilization rate* (two shifts)	52%	84%
Average metal-cutting time per day (hours)	8.3	20.2

*Ratio of actual metal-cutting time to time available for metal cutting.

SOURCE: R. Jaikumar. Post-industrial manufacturing. 1986. Harvard Business Review, Vol. 64.

TABLE 1-2 Human Resource Requirements for Metal-Cutting Operations to Make the Same Number of Identical Parts

	Conventional Systems		Flexible Manufacturing Systems
Operation	United States	Japan	Japan
Engineering	34	18	16
Manufacturing overhead	64	22	5
Fabrication	52	28	6
Assembly	44	32	16
Total workers	194	100	43

NOTE: At the time of this study, no U.S. machine tool producer had a flexible manufacturing system on line.

SOURCE: R. Jaikumar. Post-industrial manufacturing. 1986. Harvard Business Review, Vol. 64.

modes of production obsolete. Firms that hope to compete in the world market have no choice but to adopt it and learn to use it to their greatest advantage.

A 1988 Department of Defense report[2] found serious, if irregular, indications of decline in sectors of the industrial base that are critical to continued U.S. leadership in advanced technologies and, by extension, to national security. The report finds particularly devastating the erosion of production technologies and equipment in vitally important sectors such as machine tools and electronics manufacturing equipment (see Tables 1-3 and 1-4). Noting

TABLE 1-3 Top 10 Merchant Integrated Circuit Makers

Rank	1976	1986	1996
1.	Texas Instruments	NEC	IBM
2.	Fairchild	Texas Instruments	NEC
3.	Signetics	Fujitsu	Fujitsu
4.	National	Hitachi	Hitachi
5.	Intel	Motorola	Toshiba
6.	Motorola	Toshiba	Texas Instruments
7.	NEC	Philips	Matsushita
8.	GI	National	Mitsubishi
9.	RCA	Intel	Samsung
10.	Rockwell	Matsushita	Seimens

SOURCE: Microelectronic Engineering at RIT: Manpower for Tomorrow's Technology.

TABLE 1-4 Percentage of the World Semiconductor Market

Year	United States	Japan	Total (in billions of dollars)
1980	60%	27%	$10
1988	36%	53%	$46
2000	?	?	$200

SOURCE: Semiconductor Industry Association.

that these are but the leading edge of scores of other technologies in which other nations are developing advanced manufacturing technologies for advanced products, the report suggests that U.S. industry cannot hope to compete in the world market with only technological equivalence. Nor can it expect to develop advanced technology without concurrently developing a work force that is competent to use it.

CHANGING GROUND RULES OF MANUFACTURING COMPETITIVENESS

Product quality, responsiveness to customers, process control and flexibility, and development and maintenance of an organizational skill base capable of spurring constant improvement have become the determinants of global manufacturing competitiveness. Manufacturers must adapt in order to succeed in these circumstances.

This report examines five critical areas in which such adaptation must be pursued. First, to improve equipment reliability, decrease cycle times, and achieve the greater precision dictated by demands for higher quality, manufacturing will have to adopt and further develop *intelligent manufacturing control*. Second, manufacturing must maximize the productivity of its capital investments through improved *equipment reliability and maintenance* practices. Third, manufacturing must enhance product characteristics—for instance, by using *advanced engineered materials* to reduce weight, broaden service temperature capabilities, impart multifunctionality, or improve life cycle performance. Fourth, to speed time to market, manufacturing must employ *product realization* techniques and adopt organizational changes that foster effective use of these techniques. Finally, creation of the new work force—highly adaptable and possessing multidisciplinary skills of a high order—that is critical to the application of these techniques and technologies will necessitate a focus on *manufacturing skills improvement*.

Effective management of modern enterprises is characterized by a high degree of interfunctional integration and coordination at a variety of levels. In short, it requires a systems view. The new learning organization is depicted in Figure 1-5. A competitive environment imposes a continuing need for an organization to move from the present state to a more competitive state. The catalyst of change may occur at any level, but a successful transition cannot be made unless the ramifications of change are assessed and accommodated at all levels.

Advanced manufacturing technology will seldom yield the anticipated flexibility and/or productivity in an organization unless corresponding changes are made in the organization itself and in its information systems and resources. Basic changes must be made in the processes, organization, and attitudes that are common in engineering and management. In engineering, manufacturers must strive to improve interaction among design engineers, production engineers, and marketers. Managers need to reevaluate common management practices and tools, such as accounting methods, investment criteria, inter- and intrafirm cooperation, and relationships with customers, to ensure that all the firm's resources, including manufacturing, constantly are driven to improvement.[3]

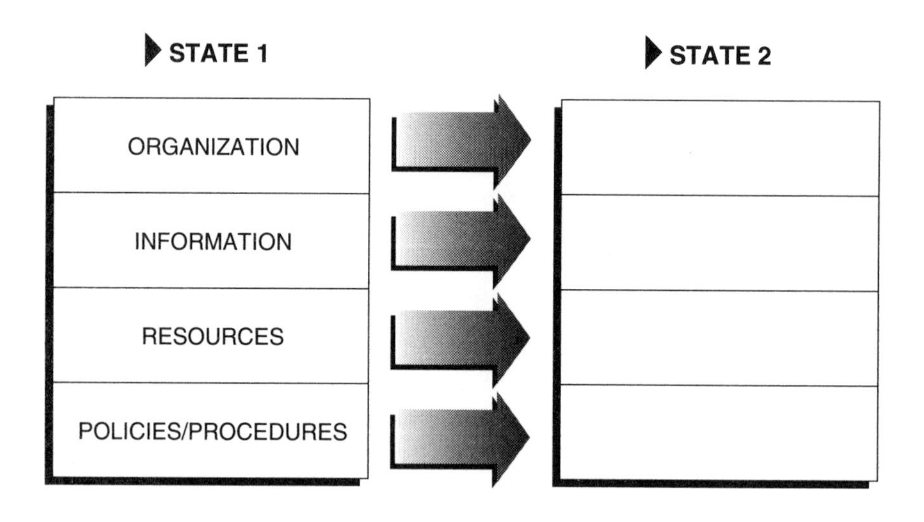

Have to Move from State 1 to State 2 – How it is Done is not as Important as Recognizing that Changing any Level will Affect other Levels.

FIGURE 1-5 The new learning organization.

Implementation of advanced manufacturing technology has a pervasive effect on engineering, management, and the whole manufacturing environment. Advanced manufacturing technology has made information and materials processing effectively one entity. Management's responsibility for directing the flow of information is diminishing as information about manufacturing efficiency and capability is increasingly captured at the factory level. Teams working in the factory can use this information to solve problems and achieve superior process capability. The vast amounts of data collected by advanced manufacturing technology enable workers to supplement experience and trial-and-error problem-solving approaches with scientific method—for example, by performing experiments and building and refining models of production. Increasingly, learning is becoming the focus of the manufacturing plant.

The rest of this overview examines conditions antecedent to establishing a research agenda for manufacturing—specifically, a three-pronged theoretical basis for manufacturing and barriers to competitiveness—and suggests steps for changing. The research agenda is developed in the subsequent chapters.

A THREE-PRONGED THEORETICAL BASIS

The changes that attend the development and deployment of advanced manufacturing technology involve the availability of *information* and the *integration* of that information with business functions to achieve various kinds of *intelligence*. Figures 1-6 through 1-10 illustrate, for each of the topics covered in subsequent chapters, representative forms of organizational intelligence that can be realized from integration of domain-specific information.

FIGURE 1-6 Organizational intelligence realized from integration of domain-specific information: Intelligent manufacturing control.

EQUIPMENT RELIABILITY AND MAINTENANCE

- Data on Economic Benefits of Improved Equipment Availability
- Equipment Requirements and Reliability Considerations
- Equipment and Process Performance Data

INFORMATION

INTELLIGENCE

- Knowledge System that Records, Categorizes, Aggregates Within Categories, and Predicts Instances of Equipment Shutdown
- Logical Model of Equipment Failure that Relates Different Kinds of, Mean Time to, and Causes of Failure and Feeds Back to Equipment Design

FIGURE 1-7 Organizational intelligence realized from integration of domain-specific information: Equipment reliability and maintenance.

ADVANCED ENGINEERED MATERIALS

- Symbolic Knowledge of Improved Materials Design
- Performance Data for Critical Parameters
- Property Data Bases

INFORMATION

INTELLIGENCE

- Knowledge-Based Systems Embodying Process Know-how and Materials Science
- Process Simulation Incorporating Broad Spectrum of Physical Phenomena
- Teaching Factory

FIGURE 1-8 Organizational intelligence realized from integration of domain-specific information: Advanced engineered materials.

PRODUCT REALIZATION

- Business/Manufacturing Strategy
- Knowledge of Product Design, Process Flow, Economic Analysis
- Knowledge of Effects of Physical Transformation
- Equipment and Process Performance Parameters
- Equipment Availability Data

INFORMATION

INTELLIGENCE

- Product Images that Provide Multiple Views to Accommodate Multiple Perspectives of Participants in the Product Life Cycle
- Systems Architectures that Provide Open, Heterogeneous Environments
- Models Capable of Monitoring Deviations in and Suboptimal Performance and Diagnosing Causes of Failure

FIGURE 1-9 Organizational intelligence realized from integration of domain-specific information: Product realization.

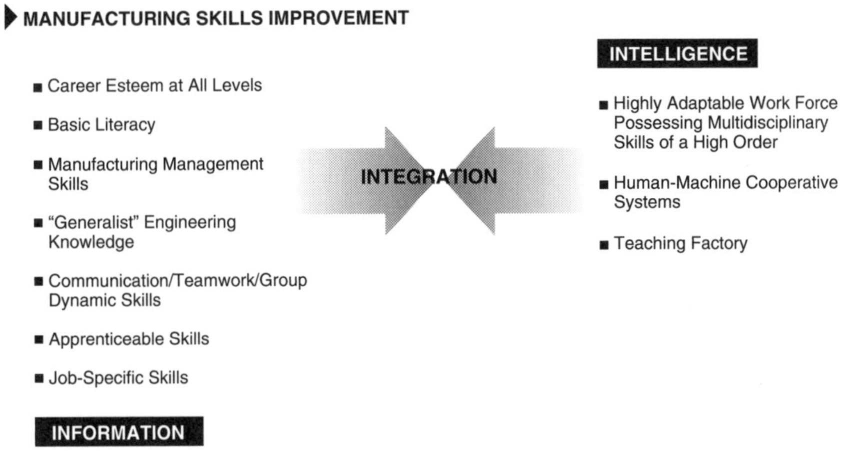

FIGURE 1-10 Organizational intelligence realized from integration of domain-specific information: Manufacturing skills improvement.

Information

Advanced manufacturing technology facilitates the collection of enormous amounts of data. The controlling computer in a computer-integrated manufacturing system records every finite state in manufacturing operations as a series of snapshots. Computers can record hundreds of thousands of these states every second; they can examine one state, control its activity by some defined procedure, and then move on to the next state. The ability of a computer to observe a phenomenon in one state and use that observation to control an activity in the next is the essence of intelligent manufacturing control (IMC).

Many computer-controlled snapshots must be aggregated to provide a picture of a time period that is sufficiently long to be useful to managers. Aggregated snapshots permit managers to compare the current state of a system to the expected state and to relate events to corrective actions. The large volume of data that such systems collect necessitates additional programming to identify and store only dynamic information that typically is part of an investigative problem-solving activity.

The enormous quantities of data include a variety of types of information. At the raw materials end are property data bases, performance data for critical parameters, and information on safety, environmental, economic, and educational factors. Equipment data range from information about what constitutes reliability and maintenance, to scheduled and actual equipment and process performance data, to data that illustrate the benefits of improved reliability and performance. Data relevant to IMC include state and historical data on the manufacturing environment as well as cause and effect data. The product realization process (PRP) relies on information on products and processes, control of equipment, and the effects of physical transformations of materials, as well as on the organization of production work, market and production problems, and variations in practice in different industries and national settings.

Integration

Information must be integrated into manufacturing operations at two levels: across process control and improvement, and across functions. Consider the integration at the raw materials end. Functional integration occurs, for example, when symbolic knowledge of improved materials design is integrated with process planning. The need to integrate measurement and control of critical process parameters with data on the performance of advanced engineered materials is addressed by IMC. Other types of integration relevant to IMC include machine and process flows in the factory, and human knowledge and machine intelligence.

With respect to reliability and performance data collected on the factory floor, integration is needed between (1) equipment user and supplier, (2) equipment requirements and reliability considerations, and (3) equipment reliability and cost accounting.

Product realization relies on integration of design and manufacturing to support parallel product development, of business logistics (to include supplier and codesigner), and of IMC with design and production. On the organizational side, the PRP relies on integration of (1) knowledge of the viability of alternative concepts, accurate evaluation of costs (e.g., equipment reliability and maintenance), market-imposed constraints, and opportunities afforded by technological change, and (2) the conception and implementation of new manufacturing technologies. (Figure 1-11 illustrates the integration of product realization, IMC, and equipment reliability and maintenance.)

Intelligence

Information is incorporated into business functions to gain a better understanding of the manufacturing process and its ele-

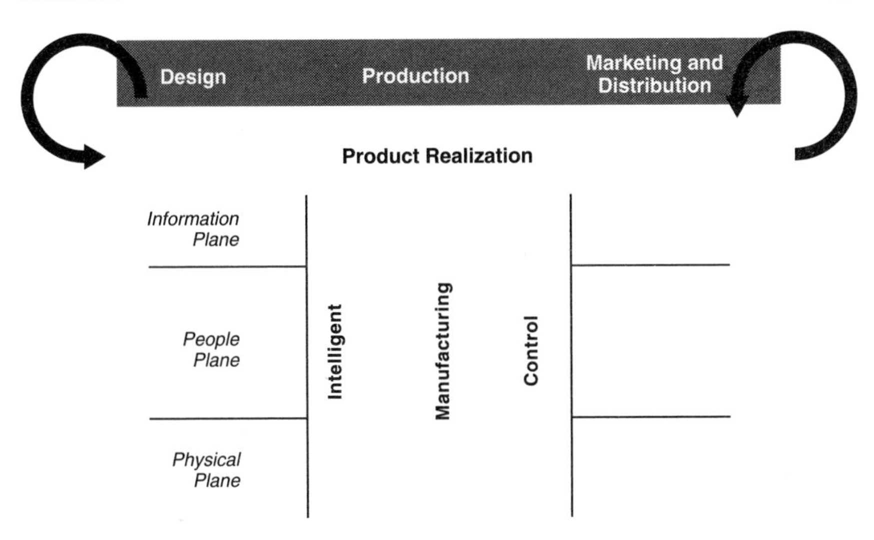

FIGURE 1-11 Integration across functional areas.

ments. The benefits of incorporating raw materials information include, for example, process simulations that incorporate extensive physical parameters and expert systems that embody both processing knowledge and materials science. Incorporation of equipment reliability and maintenance (ERM) data yields both knowledge systems that record, categorize, aggregate within categories, and predict instances of equipment shutdown; and logical models of equipment failure that relate different kinds of, mean time to, and causes of failure and uses this information in equipment design. Incorporation of the vast stores of data gathered by IMC systems facilitates the development of process models that express different views of the same circumstance at varying levels of abstraction; decision support, expert, and optimization systems; and logical models that relate disruptions to causes and are capable of learning. Pairing such developments with product realization yields product images that accommodate the multiple perspectives in the product life cycle; system architectures that provide open, heterogeneous environments; models capable of monitoring performance deviations and suboptimal performance and of diagnosing causes of failure; systems that encourage and foster nontraditional thinking; and organizational ability and incentives to be open to change.

BARRIERS TO COMPETITIVENESS

This section addresses four significant barriers to U.S. manufacturing competitiveness: inflexible organizations, inadequate technology, inappropriate performance measures, and a general lack of appreciation for the importance of manufacturing. The first three barriers may be changed from within. Industry, however, contributes to the fourth barrier by failing to provide financial and psychological incentives that might promote careers in manufacturing.

Inflexible Organizations

Manufacturing behavior in the future will continue to be driven by market as well as nonmarket factors. Market factors are those defined by technological opportunity, business strategy, and the competitive environment, and they interact to determine the investments a firm must emphasize to maintain a chosen competitive advantage. Trends in manufacturing competition and technological possibilities, for example, are driving firms to emphasize production flexibility in order to maximize their responsiveness to customers without sacrificing cost competitiveness. Nonmarket factors are outside forces such as national security, demographics, urban congestion, the regulatory regime, and social concerns over the integrity of the environment that force particular types of investments and technological developments. Nonmarket factors force firms to explore nontraditional approaches to organization and to reevaluate conventional product and process technologies. Standard approaches to solving the varied and conflicting problems that arise as a result of both market and nonmarket factors often are inadequate or completely ineffective. New manufacturing solutions, unconstrained by existing infrastructures, technologies, modes of business, or investment, are needed.

Inadequate Technology Base

The technology base of U.S. manufacturing has many types of problems, ranging from quality control to evaluation. A fundamental deficiency is the lack of standards—for design and test, for measurement and evaluation, for data communications, and for operator–equipment interfaces. Standards also are essential to the development of dynamic knowledge bases that are capable of adapting to change. Lacking as well is an information architecture that integrates human and machine intelligence to yield knowledge acquisition techniques that can support rapid start-up. Ultimately, these knowledge bases must give rise to a dynamic model that encompasses design, manufacturing, and management. Also needed is sensor integration, which will be a key ingredient of the dynamic model. Sensor integration is essential for developing reliable process models that represent the physical characteristics, structure, processing, manufacturing, performance, and cost of advanced engineered materials. And, finally, predictor maps are needed to achieve a better understanding of the interdependence of these considerations.

Such modeling and integration will rely on the development of a host of high-performance tools, including design compilers and synthesizers, rapid imaging devices, and selection analysis systems based on artificial intelligence and expert systems.

Because of the increasing importance of advanced manufacturing technology to international competitiveness, strong consideration also must be given to improving the U.S. technology skill base. This base includes both operators capable of learning to interact with intelligent equipment and managers capable of making informed decisions about acquiring and deploying such equipment.

Inappropriate Performance Measures

Much of the apparent weakness of U.S. manufacturing is attributed to reliance on inappropriate financial measures to evaluate manufacturing efficiency and corporate performance. Devised in the nineteenth century, and still the basis of the entire manufacturing infrastructure, these measures have failed to keep up with the major changes in the nation's manufacturing systems. Management analyst Peter Drucker cites four examples: (1) cost accounting systems assume that blue-collar labor accounts for 80 percent of all manufacturing costs (excluding raw materials), even though 8 to 12 percent is rapidly becoming the standard; (2) the benefits of changing a process or system are measured primarily in terms of labor cost savings (rather than in optimization of equipment utilization); (3) since only costs of production are measured, nonproduction costs, such as machine downtime or defective products, are ignored; and (4) because the factory is considered an isolated entity, only cost savings realized in the factory are significant—process changes that might increase service quality or product acceptance in the market are ignored.[4]

Alternative measures, and creative ways of substituting them for financial measures, are badly needed. In particular, measures are needed to assess the skills and levels of performance of an enterprise's employees, the relative effectiveness of its technology, the value of knowledge gained through continual refinement of its products and processes, its use and conservation of time, and concepts such as the impact of speeded cycle times or flexibility. Unfortunately, little is known about how to quantify, represent, use, and report these criteria, even though they will increasingly determine manufacturing competitiveness.

Lack of Career Esteem

The image of manufacturing as a career has not evolved at the same rate as the manufacturing environment. Despite the advanced manufacturing technology that requires operators with high-level, multidisciplinary skills, manufacturing retains a sweatshop image, characterized by dirty work in noisy environs. Manufacturing receives little promotion from career guidance counselors and the generally low esteem in which manufacturing is held is reinforced by companies' recruitment policies, which often rank manufacturing engineers below design engineers.

Even if this image problem were to be solved, manufacturing will still face a very real, and growing, shortage of qualified people. The current applicant pool from U.S. schools is largely unqualified for highly skilled, professional manufacturing jobs. The state of education in the United States is exemplified by *The Economist's* report of a sample group of 20-year-olds, of which 60 percent could not add a lunch bill or read a road map,[5] and by one employer's experience of soliciting 15,000 applicants to find a pool of 800 who could pass an elementary entrance examination.[6]

The qualifications of manufacturing management also come into question. Among Japanese users of advanced manufacturing technology, most managers have been trained as engineers. Most U.S. manufacturing managers have graduated from programs that stress financial management and have spent little or no time on the shop floor.

STEPS FOR CHANGING

The barriers to manufacturing competitiveness recounted above are not insurmountable. To use the enormous amount of information that is available to achieve integration and intelligence in the factory, however, it will be necessary (1) to restructure the organization to support learning and experimentation in the factory (the notion of the factory as laboratory), and (2) to develop new methods of performance measurement and process/life-cycle costing that will enable management to evaluate problems, process improvements, resource utilization, and production management in economic terms. Some industries are already moving forward on these fronts.

The approach to process control costing can be changed fundamentally by focusing on the cause of process or product variance and attempting to capture all of the associated economic consequences. The costing system would view the production process as running exactly as expected unless disrupted by an event, would recognize such events, identify the effects of each event on the entire production process, and report this information to management. The power of this approach lies in the ability to identify events and their economic consequences simultaneously. To relate an event to performance measures and, subsequently, to make an economic decision to control a process requires:

- the ability to translate the event and the performance measure into monetary terms,
- a scientific model that relates production parameters to process parameters,
- an economic model that relates resource utilization to process capacity,
- knowledge of all the controllable parameters and constraints on production,
- a time scale for every controllable feedback and feed-forward loop, and
- knowledge of the relationship between a set of controllables and a set of resources.

As an example of costing events, contingencies, and process improvements, consider machine downtime. It is first necessary to determine for the machine a value per unit of time. This shadow cost of capacity can be obtained through a variety of methods related to capacity utilization. A production monitoring system provides information about all machine shutdowns (e.g., causes, durations). These events are classified by category and aggregated within categories. Given the opportunity cost of downtime, it is possible to calculate the benefit of reducing or eliminating each category of downtime. The example can be taken a step further, to the development of a logical model of machine failure that relates different kinds of failure, mean times between failure, and causes of failure. As information is gathered about production processes, one can begin to assess the impact of different causes of machine failure. The model needed for process control costing is precisely the model needed to control a process. A process control costing system adds to this model the economic value and economic cost of reducing or eliminating the different causes of failure. The benefit realized equals the time saved multiplied by the opportunity cost of that time. The cost is the cost of the resources—in terms of new procedures, maintenance, new sensors or tools, and personnel—required to make a given change.

FINDINGS OF THE PANELS

Each of the five critical areas of manufacturing examined in this report is covered in a separate chapter. Summaries, including research needs, are provided below.

Intelligent Manufacturing Control

Chapter 2 examines the tight coupling of sensor technologies and software systems that manifests machine intelligence when some degree of synergy with a human interface is achieved. A framework is established for thinking about IMC in terms of domains of control that correlate with a compressed organizational hierarchy and levels of feedback time. Research efforts should be aimed at (1) developing technique-oriented communication standards; (2) refining sensor technology in data integration, pattern recognition, and actionable models; (3) building knowledge bases of design, manufacturing, and management intelligence that can adapt to changing knowledge and organizational structures; (4) creating a dynamic model of manufacturing; (5) identifying ways to use the human–machine interface to facilitate learning in an integrated environment; and (6) redefining methods to accommodate holistic research in a production environment (i.e., the factory as laboratory).

Equipment Reliability and Maintenance

ERM, covered in Chapter 3, includes both manufacturing equipment and the technical, operational, and management activities required to sustain the performance of such equipment throughout its working life. ERM has the potential to affect three key elements of manufacturing competitiveness: quality, cost, and product lead time. Several cases illustrate effective applications of ERM programs, both in the United States and abroad. The limited penetration of such programs in this country is attributed in large measure to manufacturing managers' lack of awareness of the economic benefits of improved equipment availability. Technological problems that are associated with the implementation of ERM programs are considered to be more easily solved than are people problems.

Advanced Engineered Materials

Chapter 4 focuses on manufacturing involving advanced engineered materials (AEMs). Barriers to optimization outside the normal scope of manufacturing science and engineering are considered, as are future needs and directions. Challenges to the integration of AEMs into manufacturing operations include (1) both the need to integrate processing methods into the design and development of new materials from the beginning and the need to instill awareness of and sensitivity to materials' properties into process design paradigms; and (2) deficiencies in process simulation and modeling, knowledge-based systems applications, sensor applications, and technical cost modeling. Research should focus on needs in the areas of materials science and engineering, expanded and revised educational programs and objectives, and methods for better integrating materials-specific issues in manufacturing paradigms.

Product Realization Process

Product realization is both a consequence of and a response to pressures of time-based competition. The product realization process (PRP), discussed in Chapter 5, has both technological and organizational components. Technological enablers are needed to support the development of a universal product image that will be in sync with the views of the many participants in the product life cycle. Existing information architectures will not support the development of such images. Development of a universal product image also will rely on and engender a need for alternative organizational structures. Tomorrow's business organization is expected to resemble less a hierarchy than a peer network configured for mutual benefit. Research in this area should be directed at defining, identifying specific instances of, and developing intelligent images; identifying and establishing the requisite connections among these images; and devising an organizational structure in which these concepts can be made operational.

Manufacturing Skills Improvement

Effective deployment of advanced manufacturing technology, as pointed out in Chapter 6, relies on a host of skills that U.S. schools are neither cultivating nor preparing students to acquire. The first need is basic literacy. Then, beyond primary and secondary education, is a need for general engineering knowledge and the development of communication, team, and group dynamics skills. In the workplace, greater concentration is needed on the cultivation of apprenticeable and job-specific skills. Finally, the public must be made more aware of the importance of manufacturing to the national economy, starting with the development of career guidance materials for all levels of the educational system. The new manufacturing work force, destined to function increasingly as a component of human–machine cooperative systems, must be highly adaptable and possess multidisciplinary skills of a high order.

Implicit in the research topics identified in each chapter is a need for fundamental change in both methods and kinds of research. The typical laboratory experiment is concerned with observing a piece of a system, e.g., the signal-to-noise ratio is artificially high, and many of the variables are controlled. The notion of control, taken for granted in the laboratory, is itself the object of experimentation in the factory. In a production environment, it is necessary to study an enormous amount of information, including production histories, for a variety of integrated processes. The performance of an integrated production system can only be evaluated by observing the system as a whole in the factory.

The factory as laboratory is the new research imperative. It implies new ways of doing research, new forms of collaboration across functions and engineering disciplines, and cooperation between academic scientists and industrial practitioners. Therefore, development of an architecture for learning is critical. How to sponsor and promote the needed new forms of research is a fundamental question that must be addressed.

VISION

If the research proposed in this report is undertaken and proves fruitful, if manufacturing learns from and adopts the fruits of that research, if manufacturing gains well-deserved esteem, and if the nation proves equal to the task of redirecting and reinvigorating its educational system, manufacturing might achieve the vision summarized here.

Next-generation manufacturing will add more value during fabrication by placing greater emphasis on tailoring the specific properties of materials to specific uses. Highly specialized processing of metals, ceramics, and polymers stacked in interpenetrating layers will yield radical improvements in materials' strength and toughness and greatly reduce weight. Techniques for producing semiconductor materials at lower temperatures will dramatically reduce defects in devices of extremely small geometry. The design of next-generation manufacturing equipment will occur along with the design of the processes that will use this equipment and will draw on extensive bodies of performance data and user experience. Recognizing the productivity gains that can be realized from improved equipment availability, manufacturing managers will implement extensive programs of predictive maintenance. Maintenance will be performed by equipment operators, interacting synergistically with intelligent control systems capable of predicting and, in some instances, preventing or automatically remedying, equipment and control systems failures.

Intelligent control will extend throughout the manufacturing system. Autonomous control systems that achieve synergy between human and machine will exploit human powers of perception, pattern recognition, and problem solving in conjunction with machine capacity for manipulating vast amounts of quantitative data to learn from each situation encountered and decision made. The plant will become a locus of learning; linkages among individual intelligent controllers will support a systemwide view of the interrelationships among unit operations in the manufacturing cycle, facilitating the timely sharing of important process revisions, quality information, and overall system objectives.

Product realization techniques will increasingly substitute the content of a burgeoning knowledge base for traditional material inputs to the manufacturing process. The reduction of product and process specifications and manufacturing capabilities to intelligent images, capable of interacting with one another, will permit entire product life cycles to be simulated to evaluate tradeoffs before producing a prototype. Indeed, prototypes will increasingly become the first units of production rather than preproduction models. Emphasis on maximizing capacity utilization and minimizing investment will shift production to a largely subcontracted function. Firms, radically restructured internally, will use advanced communications technologies to manage external relationships in a constantly shifting pattern, often cooperating and competing on different contracts at the same time. Global information systems will enable electronic virtual enterprises to access and coordinate the localized design and manufacturing capabilities of village industries all over the world. Nations' competitive positions will be determined by their positions in the world information market, which in turn will be determined by investment in the information infrastructure. World boundaries will be determined less by national affiliation than by class affinity for information categories.

Production will increasingly be handled by highly skilled professionals functioning as components of human–machine cooperative systems, and integrally linked with management functions in a compressed ("flat") organizational hierarchy that is characterized by peer-to-peer relationships. Manufacturing careers will be highly regarded and avidly pursued by well-educated youth who have been familiar with manufacturing enterprise since elementary school. Colleges will turn out Renaissance engineers and business schools managers who understand technology and the workings of their plants as well as the composition of their balance sheets. Job-specific skills will be imparted through nationally coordinated vocational and apprenticeship programs and satellite distribution networks.

Manufacturing will be regarded as a national asset, to be protected, cherished, and nurtured. Esteem for manufacturing will rise to the level of its technology.

NOTES

1. Manufacturing in this report is understood to be the processing of raw material inputs and the assembly, mixing, or other coalescence of outputs into high-quality, low-cost, salable products. The contrasting use of "manufacturing" in Chapter 6, Manufacturing Skills Improvement, is not a definition but a very important image issue. It reflects a perception of the definition that is increasingly at variance with the reality of manufacturing that employs advanced technologies. Qualifying this perception is very important to building career esteem.

2. Under Secretary of Defense (Acquisition). 1988. Bolstering Defense Industrial Competitiveness. Report to the Secretary of Defense. 1-2, 6-10.

3. Hayes, R. H. and Jaikumar, R. 1988. Manufacturing's crisis: New technologies, obsolete organizations. Harvard Business Review (September-October): 77-85.

4. Drucker, P. F. 1990. The emerging theory of manufacturing. Harvard Business Review (May-June): 94-102.

5. Gone fishing. 1990. The Economist. 314 (January 6): 61-62.

6. Richards, B. 1990. Wanting workers. Wall Street Journal Reports: Education (supplement) (February 9) R10.

2

Intelligent Manufacturing Control

MANUFACTURING CONTROL historically has been adaptive, using sensors to detect out-of-tolerance conditions, feeding the information to a controller, and changing process parameters to bring output back within tolerance limits. This highly localized approach is no longer sufficient. As processes grow in complexity and as intense, increasingly global competition drives firms more frequently to introduce products with more variations, the need to augment existing process control techniques has grown apace. This chapter describes the tight coupling of sensor technologies and microprocessor-based software systems that manifests intelligence by learning from experience and exhibiting some degree of synergy with a human interface. Here, we present a framework for thinking about *intelligent manufacturing control* (IMC) in terms of a compressed organizational hierarchy and shorter feedback time.

IMC is a distributed, hierarchical approach to the control of manufacturing processes. It employs electrically coupled computer-based hardware controllers and process sensors in conjunction with a trained, self-directed work force to process physical state and historical data derived from the manufacturing environment. IMC has a twofold objective: (1) to satisfy product quality and process control requirements for existing products and processes, and (2) to be adaptable enough to do the same for future products and processes by providing a way not only to control the manufacturing process, but also to promote learning that will lead to process improvement.

In a broader context, IMC encompasses all systems that affect the manufacturing floor, including:

- product and process design systems, including engineering and vendors;
- facility systems, including environmental and maintenance support;
- personnel systems, including training and subcontracting;
- order entry and requirements forecasting by sales and marketing personnel, dealers, and distributors; and
- physical distribution systems, including warehousing and transportation.

In practice, these systems usually are considered only at the interfaces to the manufacturing environment, where their effects are described heuristically and statistically.

To describe IMC in the broadest context, five levels of traditional plant hierarchy are here compressed into three, which permits a simpler organization and provides an avenue for establishing interactive links with manufacturing areas examined in other chapters. This compression is consistent with the current trend in industry toward a general flattening of organizational hierarchies.[1] The requirements for IMC include a temporal dimension.

Figure 2-1 uses a logarithmic scale to correlate 10 levels of feedback time—ranging from 0.01 second to one quarter of a year—with three domains of intelligent control. These domains are shown opposite the three levels of the factory hierarchy on the vertical axis.

In the domain of process control, a precisely stated contingency procedure operates in real time at the machine level without human intervention. In the domain of observation and pattern recognition, the efficacy of procedures defined in the domain of process control is observed; contingencies in the behavior of procedure are studied; and improvements are made. Problems are solved at the cell level. The domain of learning and improvement is one of choice, where the options available for improving a system are assumed to be numerous and available resources to be limited. It is in this third domain, at the plant level, that economic choices are made about which avenues of process improvement to pursue in view of supply and demand, resource utilization, and other production management functions. IMC spans all three domains.

Finally, this chapter assesses the human skills and machine complements that are needed to achieve IMC. In general, high-speed technologies tend to push decision making down to the

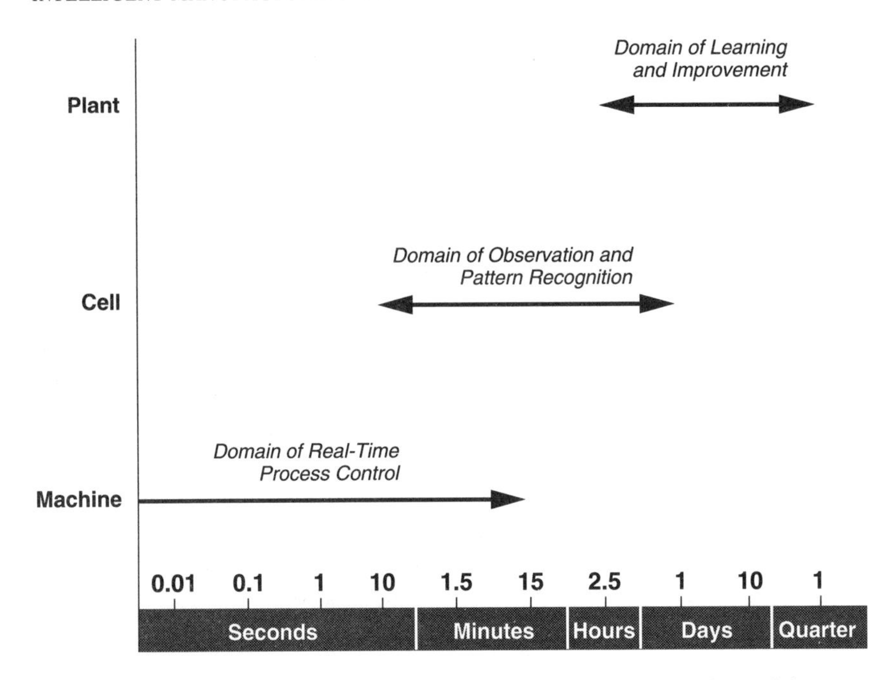

FIGURE 2-1 Levels of control versus hierarchy. Source: Adapted from R. Jaikumar, An architecture for a process control costing system, in Measures for Manufacturing Excellence. R. S. Kaplan, ed. 1990. Boston, Mass.: HBS Press. 193-222.

unit level. Increased decision making at this level requires more highly skilled and autonomous workers, and more effective computer-based, real-time control diagnosis and decision support tools. (See case studies in the section on present and future practice on pages 42-49.)

Current sensor technology encompasses visual, ultrasonic, thermal, chemical, inertial, electrical, tactile, and audio sensors. These can be used singly or in combination to

- provide highly detailed macroscopic information on dimension, position in space, shape, velocity and acceleration, global or local temperature, and compositional distribution;
- detect a variety of other physical, chemical, electrical, optical, and magnetic properties;
- probe internal macro- and microstructure to measure parameters such as grain size, texture, and the presence, size, and distribution of voids or other defects; and/or

• determine material characteristics at the atomic and molecular levels.

Used as transducers, to trigger as well as sense signals, and to perform analog-to-digital conversions, such sensors can monitor and control a wide range of manufacturing operations and processes. IMC is concerned with using these devices in concert with human knowledge and automatically learned relationships to provide closed loop, real-time control of critical engineering and manufacturing processes. This control is essential for process stability and for maximizing quality, performance, and reliability while keeping costs low.

In sum, IMC implies an ability to (1) assimilate and validate sensory information from a variety of sources, (2) make reasonable assumptions about an operating environment, and (3) execute suitable action plans based on both scientific models of a process and experience gained from executing prior action plans. A model that can learn from prior actions and adjust itself accordingly must thus be devised.

IMPORTANCE

IMC has the earmarks of a science that today is uncharted, but that has tremendous implications for the future. Its extensive use of computer technology builds on U.S. strengths in cognitive science and computers and systems and provides an avenue for strengthening the transfer of knowledge from the laboratory to the manufacturing floor. This will occur as the distinction between laboratory and factory fades, with the latter necessarily becoming the locus of experimentation. IMC will change the way people think about knowledge transfer. It will change the culture of improvement, shifting emphasis away from transferring technology from the laboratory to creating and exploiting technology in the factory.

The need for IMC systems is being driven by the increased precision and decreased cycle times demanded by today's intense competition and by business needs for improved product quality. Given the increasing use of technology in manufacturing, and the growing volume and complexity of information and information sources, unaided human decision making is becoming less and less optimal—decisions made by people simply take too long and fail to reflect the richness of available data. But to leapfrog to the concept of the "lights-out" factory is to overlook the value of the people on the floor. In order to process increasingly complex information in a shorter time, new tools must be developed to leverage human cognitive abilities. This is precisely the vision for IMC.

Even within the limited context of the manufacturing floor, the issues associated with IMC are numerous and technically challenging. Yet to be resolved are issues faced by early users and suppliers of such systems—issues related to data acquisition, correlation, presentation, and quality control, simulation of control decisions, learning from process disruptions, understanding process complexity, standardization in system implementation, and more efficient user training. A description of each of these issues follows.

Data acquisition. U.S. manufacturing is characterized by thousands of types of data, interfaces, and sensor requirements. One issue is the infrequent use of computerized machine controllers by U.S. manufacturers relative to manufacturers in other countries. Without such controllers, data for IMC are simply not available. Another issue in process measurement is the low reliability of gauges, which can corrupt control data. A third issue is the need for communication protocols for moving data to control points.

Data correlation. The problem of combining different types of data must be solved. When the underlying process physics are understood (e.g., in steel rolling), data fusion can be algorithmically described and handled by computer-based controllers. In most manufacturing today (both discrete parts and process industry), data fusion is performed by a human operator or supervisor.

Data presentation. Presentation of data becomes critical when humans act as controllers. X-R charts for statistical data are one approach—well-known in manufacturing—for reducing the time to operator action.[2] More such presentation methods are needed to reduce to actionable information the enormous amount of available manufacturing data.

Data quality control. Collection and correlation of data is necessary, but not sufficient, to meet the needs of IMC. Data from which knowledge is to be extracted must be of uniformly high quality; this is generally not true of raw data. Automatic methods of identifying and eliminating errors, gaps, and redundancies in data are needed.

Simulation of control decisions. Data presented to a human controller must be analyzed and a control option selected. To optimize the control decision, historical information in the form of expert advice or root cause analysis is required. It should be possible to simulate a critical control decision to ensure that it is the optimum selection.

Learning from process disruptions. Each disruption of a process needs to be recorded, the problem identified, its cause determined, and a way devised to prevent it from happening again. This task requires access to processing data that may be highly precise, at the microsecond interval and across the entire system, and include historical data on similar disruptions. In addition, such data, being problem-specific, would change from one disruption to another. Process learning relies on an ability to adjust rapidly to the changing needs of different problems and to link information from a wide variety of data bases. To be useful, such data must be statistically significant and thus may require variable recording times.

Understanding process complexity. In certain industries, such as the integrated circuit industry, the complexity and number of processes overwhelm all other considerations. In some chip manufacturing situations, for example, the process limits are so strictly adhered to that a yield of only a few percent is allowable. In such industries, the basic need is for a much better understanding of the processes involved. A scientific point of view might dictate explicit modeling of the processes. From a control perspective, it might be enough to understand the interaction parameters sufficiently well to provide appropriate control compensation in real time.

Standardization in system implementation. Because every manufacturing environment is unique, standardization of control architectures, communications, data base structures, information graphics, and computer applications software is necessary to reduce the cost of implementing IMC. Creative financial programs and approaches to technology upgrades are needed to encourage start-up.

More efficient *user training.* Because control is distributed and centralized decision making is often inefficient, IMC systems require a new type of operator and new organizational structure. The work force must be trained to use computer-based information tools to make local control decisions. Management must be reorganized and its role changed from traditional decision making to coaching. Managers, particularly in smaller companies, must be shown that new manufacturing technologies not only are available, but also are vital to their firms' long-term health.

VISION

To imagine manufacturing with intelligent control, it is helpful to recall the Martian Rover. Deposited on the surface of Mars, this highly reliable mobile robot manipulated a variety of tools, responded to numerous sensors, performed experiments, and learned from and adapted to its environment. Its use of hierarchical, local, decentralized control to operate untended and deal with contingencies in an unpredictable environment was observable remotely (from another planet!), allowing its control algorithms to be studied, improved, and effectively reintroduced into the environment. The notion of remote control—of running a process from behind a wall, without seeing, touching, or feeling any part of it—is implicit in IMC.

The human observer brings to the analysis of an unfamiliar scene broad knowledge tied to perceptual skills, and the ability to review, and perhaps rethink, a scene from different perspectives. Such dynamic interplay among sensors, analysis, and knowledge is necessary for learning about a sensed environment to be able to respond to changes in it. Lacking this integration, current machine vision and sensor systems have limited ability to work in uncontrolled or changing environments. Many plant installations of sensor technology have ultimately failed, not because the sensor technology was wanting, but because conditions changed relative to those for which the sensor and control strategy were optimized.

Current inspection and robotic guidance analysis systems must accept input in whatever form sensors provide it. The sensors do not know whether the data they provide is of any use and the analysis system does not know what specific changes in the data might mean (e.g., that a light bulb has just burnt out). The basic dynamic interplay employed by a learning child can close the loop around data acquisition and analysis; the development of computer-based analogs, such as the Cooperative Hierarchical Image Learning Dynamics (CHILD) begun and partially completed by the Industrial Technology Institute, could be a major step toward truly autonomous robotic capability.[3] It is this dynamic interplay that leads to adaptation and modification of knowledge, which, in turn, opens the way to novel inquiries and greater understanding. Application of such image learning dynamics would produce truly flexible manufacturing systems that could identify or sort mixed parts or direct robotic applications in less controlled environments.

CHILD entails development of four basic building blocks as a foundation for the required interplay of dynamic interaction. These are:

- an adjustable image acquisition module, which includes adaptable imaging, lighting (irradiance), and/or any other sensor systems to be used;

- an image (or other data) analysis module, which will be the computer complement most suited to analyzing the data provided and controlling the parameters at hand;
- a knowledge base module, which will make a broad base of knowledge—of the sensor systems and what they can do, of the environment and natural laws, and of the object (be it a production part, a machine, or a satellite)—available on a hierarchical level; and
- an executive control module, which will be able to solve simple problems, learn from experience, and coordinate the activities of the other modules.

Learning about the environment and effectively responding to contingencies involves a synthesis between human and machine, local control, and central intelligence. An intelligent controller must be able to communicate with heterogeneous data bases, learn from similar or related instances, and incorporate a model of control for different contingencies. This model must be autonomous, i.e., capable of operating with or without human intervention. An even more powerful paradigm for control exploits the human powers of perception, pattern recognition, and problem solving and the intelligent manufacturing system's ability to manipulate vast quantities of procedural knowledge. Manufacturing control systems built on this paradigm exhibit synergy between human and machine and an ability of both to learn and they incorporate a dynamic model of this world.

The vision for intelligent control is of control across the breadth of the nation's manufacturing systems. It is not implementing optimal control for achieving stated processing goals for a unit operation, but rather creating systemwide views of processing that take into account the interrelationships among individual unit operations in the manufacturing cycle. Given such views, it is possible to affirm the goals for any one of a chain of processes and clearly recognize the implications of material, process, or product deviations. With linkages among individual intelligent controllers in the manufacturing chain, important process revisions, quality information, and overall system objectives can be shared with minimal lag time.

The intelligent controller implicit in this vision must be capable of establishing and executing process plans that both reflect operator know-how and model-derived principles and are aimed specifically at controlling product attributes. To maintain processing flexibility in an environment characterized by rapid maturation of process, material, and product applications, IMC must exploit control technologies that do not (1) restrictively prejudge interrelationships among key process parameters, (2) assume certainty of control objectives, or (3) prematurely determine linkages between process behavior and product quality.

Adaptive control, though it responds to environmental change, is based on some fixed model of a process and is local in terms of the information it gathers and the control it exerts. IMC, in contrast, analyzes and uses historical information about its own actions, together with systemwide information from many sensors, to adjust its model of the world and effect novel action plans.

Knowledge-based systems that clone the knowledge of one or more experts to improve understanding and control of a process already are in use in some applications such as steam turbine generators, diesel locomotives, and wave-soldering machines.[4] The next steps are to effect a synergy between these expert systems and the humans whose knowledge they embody and to implement methods for learning by example. This will optimally be achieved by using all available information. This information might take the form of known rules and algorithms or patterns of response learned by example. Learning must take place with every decision and situation. The characteristics that make manufacturing environments ideal for learning—information sources that are multiple, complex, dynamic, and accessible—are also responsible for the difficulty of learning.

IMC is more than knowledge-based systems for process control; its purpose is not to exploit operating expertise, but to convert operating experience into a manufacturing science. This conversion will require very close interaction between the scientific community and the manufacturing plant. The plant must be the locus of research because only there can entire processes, rather than isolated steps of a process, be studied. At the same time, it is the scientific process that provides the methods of learning by which a logical model capable of representing entire processes can be built. In plants, pressure for control overwhelms scientific understanding. The scientific community's interest in the development of a manufacturing science must be turned into the resources required to build needed learning systems within the context of the plant.

PRESENT AND FUTURE PRACTICE

The implementation of IMC depends on both machine-related and business- or environment-related manufacturing assumptions, greater intelligence in the form of precise and complete sets of contingent procedures, and versatility and generalization of deep process knowledge. It relies on the replacement of scale economies by learning economies and the existence of a systemwide architecture that encompasses information structure, organization of human resources, and process structure at all levels of the hierarchy and on the nature of the decision making at each level.

Manufacturing Assumptions

Manufacturing assumptions today are vastly different from those of only a few decades ago. A fundamental shift in the paradigm of production is taking place—from managing materials processing to managing information—in which machines are seen increasingly as extensions of the human mind. Any discussion of IMC, whether in discrete or process industries, must consider two broad sets of manufacturing assumptions, machine-related and business- or environment-related.

Modern systems characterized by integration and intelligence are appropriately viewed as human-machine cooperatives. To understand the significance of this shift, imagine the technology in the extreme. Consider a small group of engineers using a connected system of workstations to design and write the software for producing a product on any defined configuration of equipment anywhere in the world. Once the procedures are created, machine capacity and materials become commodities to be bought and sold at whatever price can be obtained. What is preventing such a development is not the absence of mechanization, but the need for greater intelligence in the form of precise and complete sets of contingent procedures. Developing these intellectual assets is today's technological imperative.[5]

The development of intellectual assets rests on a fundamental, machine-related assumption—the versatility and generalization of deep process knowledge. Newer manufacturing systems have distinct advantages: they can be used to produce many different products, they are adaptable to changes in design or recipe, and they can operate untended. Investment in such systems—in people, equipment, and software—must be made before production begins. The versatility of manufacturing systems and transportability of procedures make the market for production capacity highly competitive and subsequently make the capacity itself a commodity. The only premium that can be extracted lies in the creation of new procedures for improved processes and products.

Scale economies are replaced by learning economies, and firms compete by trying to create performance advantages and introduce new products more quickly.

The versatility and generalization of deep process knowledge rests on another machine-related assumption—the existence of a systemwide architecture that encompasses information structure, organization of human resources, and process structure at the plant, cell, and machine levels.[6] The architecture of factory work that integrates information and materials processing, automated auxiliary functions, and extreme flexibility and versatility would probably be hierarchical. The system would use specialized, dedicated computers to operate machines, material-handling facilities, and inspection processes and to manage cells throughout the plant in an integrated manner. Such hierarchical organization is used in almost all existing automated manufacturing systems and corresponds to the vertical organization of most present-day factories.

Levels of decision making associated with these factories also are present in the hierarchical computer system (see Figure 2-1). At the lowest level, concerned with actual machine operation, is process information. Decision making at this level can employ adaptive control. At the intermediate level, where manufacturing operations are managed and contingencies and conflicts are resolved, are cell controllers. The principal decision-making function at this level is coordination and management of resources. At the plant level, the principal decision-making function is management of experimental procedures, capacity, and knowledge bases, and the resource allocation methods for learning and control are strategic.

Other manufacturing assumptions related to the machinery of production are high reliability, failure recovery (graceful degradation and restart), the existence of workstations and massive amounts of information, and consistency of data.

The fundamental business-related assumption of the new manufacturing environment is an economic representation of a world model, a complete description of contingent procedures for manufacturing control viewed from the perspectives of factory, product, and process. These descriptions may reflect different levels of abstraction and precision, but must be consistent across the hierarchy.

The increasingly innovation-based nature of competition demands quick start-up and transition with minimum waste of materials, time, and human resources. This situation gives rise to business-related manufacturing assumptions associated with the development and management of intellectual assets, including:

person–machine interaction (operator training and instruction, and ease of use in a more complex world), economic management of disruptions (a new architecture for process control costing), explicit modeling of disruptions, and intergenerational learning.

Contingencies are typically brought to light by a product's failure to conform to specifications. Discrepancies—whether in consistency, location, shape, size, surface finish, or volume—can result from combinations of changes in four broad categories: mechanical, thermal, operational, and feedstock properties. Also of concern is whether changes are systematic (occurring with approximately the same magnitude each time), random (occurring each time with different magnitudes and without apparent pattern), or both. Recognizing, diagnosing, and learning from contingencies requires high technical intelligence, both human and machine, and grounding in the scientific method.

In the following section, a model for intelligent manufacturing control is developed, and the consequences of the various machine- and business-related manufacturing assumptions in the areas of integration, control, and IMC.

An Architecture for IMC—the World Model

Figure 2-2 shows a classical model of adaptive control. Ideally, inputs are fed into a black box and subjected to procedures that produce an expected output. Environmental effects on the procedures may result in an actual output that differs from the expected output. Adaptive control consists of adjusting procedures to compensate for the detected difference between actual and expected outputs, thus moving actual output closer to expected output. Adaptive control assumes that differences between actual and expected outputs arise from environmental conditions and are therefore random; its only response is to make whatever adjustments are necessary to get back to the target value.

A simple example of adaptive control is thermostatic control of room temperature. If the temperature rises above the upper setpoint on the thermostat, the heat is shut off; if it falls below the lower setpoint, the heat is turned on. The system cannot discern external causes of variation, such as the window being left open. It simply continues to turn the heat on and off when the temperature falls below or rises above the established setpoints. Beyond certain tolerances, differences between actual and expected outcomes become disruptions. IMC views these disruptions as having both random and systematic elements, and treats the latter as

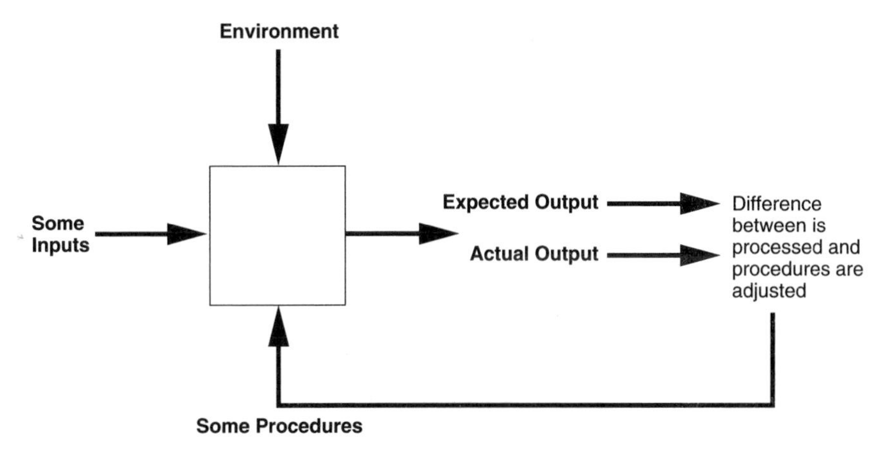

FIGURE 2-2 Classical model of adaptive control.

assignable to a cause. Returning to the example of thermostatic control, an intelligent system would attempt to identify a systematic element at work in too-rapid fluctuations in temperature. To do so, the system would need a logical model capable of representing cause and effect relationships associated with the disruption. This model might consider, for example, outside temperature, building insulation, and inside temperature and incorporate heat transfer equations and a mechanism for controlling doors and windows. Because it must associate a cause and effect relationship with every disruption, and because disruptions are evolutionary, such a model can never be complete. It is thus always a model of search with a virtual structure.

The logical model supplies the structure needed to relate a disruption to its possible causes and to refine the model progressively as more is learned about the process. Beginning with a simple matrix of causes and effects, progressive learning can lead to a scientific/mathematical model that can predict from a change in one parameter its consequences for downstream processes.

Figure 2-3 shows the architecture of a world model for IMC. It has a process control loop for each process step and a logical model for every disruption that can occur at that step. This part of the world model extends the simple input/output model of adaptive control to take into account all known parameters of a process step and to deal with disruptions that occur over a number of process steps.

The progression from one logical model to the next is the

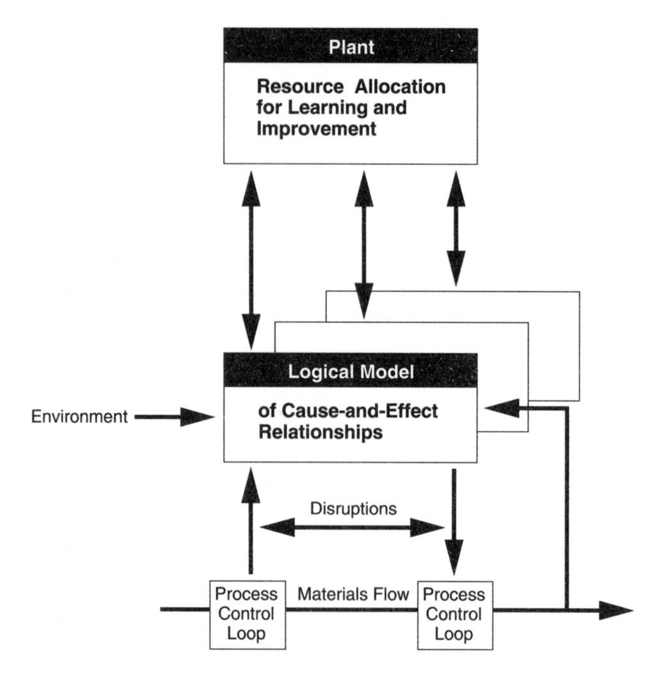

FIGURE 2-3 The world model for IMC.

essence of IMC. At different stages of knowledge about a process, one would require different information and might run different experiments. These experiments may be explicit of or implicit in the operation of the process.

At the plant level, manufacturing control consists in systematically choosing which disruptions to address. Here, an economic representation of the effects of disruptions is required to guide product and process choices for the plant. In the world model, this representation takes the form of a logical model for deciding what resources to allocate for problem solving and learning. Because the factory is a dynamic, uncertain world, the world model is always evolving. Disruptions provide opportunities for learning, which in turn require the commitment of people and money. Modern, microprocessor-based manufacturing systems make available an enormous amount of disaggregated data. IMC supports the development of an evolutionary world model by allowing these data to be organized along the domains of process control identified earlier.

The following sections examine the world model in terms of integration, control, and intelligence.

Integration

Integration in IMC involves data accumulated over time, including data on past disruptions, and implies the ability to relate current disruptions to earlier, similar disruptions. In this temporal context, IMC exists at three levels: (1) between machines and the flow of processes within a factory; (2) among different functions, such as design, engineering, and manufacturing; and (3) between human knowledge and machine intelligence.

Machine–machine integration relies on an information structure that is capable of supporting the breakdown of tasks in an uncertain, dynamic environment. Necessary are standards for data communication to support the comprehensive information objectives of the firm and a consistent world model that represents the factory as a system. The latter requires that performance levels be consistent across the hierarchy of control from plant to machine.

Functional integration implies that information from the factory can be used as a basis for learning, improving processes, and assisting in the design of the next generation of products and processes. Process models need to be constructed that can express different levels of abstraction and different views of the same circumstances (i.e., the manufacturing view, the product view, and the process view). Such models of design and manufacturing management must be consistent across all three views and be able to adapt to change.

The third level of integration, between human knowledge and machine intelligence, is discussed in the section on intelligence.

Control

Process control in a manufacturing plant relates to how a product is made rather than to what is made or when it is made. Process control generally takes place at the machine and the system levels. At both levels, control may be open loop, which requires one to recognize patterns and take action, or closed loop, which removes the human decision maker. The typical view of control presumes the existence of sensors that measure process outcomes, a model that recognizes discrepancies between expected and actual outcomes, and an algorithm that determines which process parameters must be changed and how (either to feed forward to correct the process in subsequent steps or feed back to correct subsequent output, or both).

In a noisy world with faulty sensors, incomplete knowledge

of cause and effect relationships between process parameters and outcomes, and dynamic changes in the environment, closed-loop control is difficult at best. Even at the machine level it is not very effective. Control theory has provided some impressive results, but these generally have been applied to very restrictive domains. This situation is changing, however, with the growing ability to gather, assimilate, and make sense of masses of information in a variety of forms using developing techniques in artificial intelligence and with the building of intelligent systems that exploit available information and synergies between people and machines.

Intelligence

To use the flexibility inherent in modern manufacturing systems competitively, the design process must be speeded significantly. The rapid learning implicit in doing so can be facilitated by using machine intelligence as an adjunct to human knowledge. Intelligent systems foster such synergy.

The criteria for an intelligent system are that it be used for learning, that it be focused on technological know-how, and that its intelligence be the joint product of a person and a machine. This definition encompasses a variety of systems available today, including decision support, expert, and optimization systems.

To exploit aspects of cognitive science that can lead to a technology of problem solving—the foundations of an applied cognitive science—it is not enough to model the mind as a machine, as in early artificial intelligence research, or to replicate human problem-solving processes in computer programs, as is done in expert systems. Instead, the knowledge of a given problem domain must be exploited by (1) separating well understood or formal elements from poorly understood or informal elements, (2) using the formal elements to enhance understanding of the informal elements, and (3) continuously transforming the informal elements into formal ones.

The separation and subsequent reintegration of formal and informal elements is the essence of cognitive activity and the function of any intelligent system. As the progress of technology is marked by an increase in formal abstraction, the architecture of an intelligent system can be judged by its inherent degree of formality and the extent to which this can be increased. To understand how formality is enhanced in an intelligent system, the problem-solving process—how problems are recognized, posed, and solved—must be studied.

How are problems recognized in technological development? The traditional view is that they are recognized in the design stage and steps are taken to forestall them. In practice, problems are more often recognized after the fact, as contingencies arise and causes are sought. Problems can only be recognized at the design stage in manufacturing environments that are well understood, which implies that they are static. More generally, problems arise unexpectedly in uncertain, ambiguous, and dynamic environments.

How problems are posed is inextricably linked to the way knowledge of a problem is organized and represented. The traditional view is that knowledge is organized in categories, and that theories are developed that relate these categories, creating new ones or collapsing several into one. The dynamic view is that knowledge is organized around prototypes that have implicit internal relationships, and that problems are posed as searches aimed at identifying the degree of similarity to a typical member of a prototype and elaborating on the internal structure of that prototype.

How are problems solved? Here, again, the extreme views can be characterized as traditional and dynamic. The traditional view holds that precise procedures can be written for solving problems through logic and reasoning. The dynamic view is that problems are solved by means of some combination of experience, judgment, experimentation, intuition, and skill.

These views have a certain consistency. The traditional approach to problem recognition, the organization of knowledge into categories, and the formal method of problem solving are consistent with the notion of technology as science. The dynamic approach to problem recognition, the organization of knowledge around prototypes, and informal methods of problem solving are consistent with the notion of technology as expertise. These two perspectives lie at the upper and lower bounds of process knowledge. An intelligent system can simultaneously take both perspectives and exploit the synergy between science and expertise to move progressively to higher planes of knowledge.

An architecture of control for an intelligent system relies on five central premises:

1. problem solving is begun with partial knowledge of the problem domain;
2. this knowledge comes in chunks;
3. these chunks can be formally represented and manipulated;
4. relationships between chunks can be seen, theorized about, and tested in the external environment; and

5. human rationality is bounded and judgment theory is value-laden and biased.

Case Studies

Two cases—one involving a wire-drawing operation, the other a chemical plant—are used here to illustrate issues of control, integration, and intelligence in IMC.

Both cases involve numerous processes that are subject to disruptions. Because a disruption in one process can originate in an earlier process, IMC must be able to establish such relationships. Traditionally, this has been done by searching historical data for "similar" discrepancies between recipes and effects. The concept of the logical cell (introduced with Figure 2-1, and expanded upon in the wire-drawing case and implicit in the chemicals case) provides a structure for searching process control data bases and for controlling process parameters through closed-loop feedback. The objective is ultimately to move from a rule-based system to a system that begins to approach a science. IMC at this level serves the development of recipes and products, as seen in both of the following cases.

It is readily apparent from these cases that much work is needed to achieve the vision of IMC described in this report. The state of the art will not move beyond that depicted in these cases unless industry and academe mount a concerted effort and commit the necessary resources.

IMC in the Wire-Drawing Industry

The 1985 introduction of microprocessor control in the wire-drawing industry and that industry's rapid adoption of computer-integrated manufacturing (CIM) have afforded an opportunity for cooperative development of an architecture for IMC. Preliminary results of the efforts of one firm working jointly with academic researchers to introduce IMC in one of its wire-drawing plants are reported below.

To preserve the firm's anonymity, this account is set in a generic, pre-1985 installation. A typical wire-drawing plant has two large pickling machines, 200 dry wire-drawing machines, 20 heat-treating installations, 1,000 wet wire-drawing machines, and 100 finishing lines, all laid out functionally. Such a plant makes between 150 and 1,000 different products. Wire rod received on spools is tested for properties related to the raw material and stored. The spools are subsequently pickled (cleaned in an acid bath), then loaded into an unwinding spool bin and the wire is pulled through a series of progressively smaller dies to lengthen it and reduce its diameter by adding stress and changing its crystalline structure. Next, the wire is heat treated to relieve the stress and then coated with different substances to change its surface structure. The wire is wound back onto spools, transported to a wet wire-drawing operation, and then to finishing operations such as cutting, galvanizing, and coating with adhesives.

Wire fractures, the most vexing problem in the wire-drawing process, result from process variance and are reflected in poor-quality end products. Process variables number in the hundreds, as do possible responses to a wire fracture. To enable operators to cope with these many degrees of freedom, attention-focusing and control mechanisms are needed.

Today, different steps in the process are located in different parts of the plant under their own supervisory structures. Information about the impact of heat treating, which is done in one part of the plant, on wire-drawing, which is done in another, is not captured. In fact, process variance is not analyzed systematically; a process that is out of control is handled by ad hoc engineering analysis. Consequently, no learning occurs and history repeats itself.

IMC could provide the mechanisms needed to identify problems and adjust process parameters, both upstream to prevent problems from recurring, and downstream to correct for process deficiencies. It could also help management make trade-offs between managing production and running experiments to isolate process problems.

An automated, continuous wire-drawing process is currently being developed that places process flows under the control of one system for similar products, significantly reducing the required machine complement. The typical automated plant will probably still have two pickling machines, but only 10 dry-wire drawing machines, one heat-treating installation, 500 wet wire-drawing machines, and 50 finishing lines (representing a 95-percent decrease in dry-wire drawing and heat-treating equipment, and a 50-percent decrease in wet-wire drawing machines and finishing lines).

When all of these microprocessor-controlled machines and the processes that run on them are integrated under a hierarchical control structure a decision maker will be able to track every process parameter that operates on every meter of wire that goes through the line. For example, during heat treating, the location of a given meter of wire can be known when different furnace burners are opened and closed. Thus, the decision maker can know not only the ambient temperature of the furnace at any given time, but also the finite states of a number of operating parameters. This means that the decision maker will have data, never before available, that describe the sequence of events that acted on a meter of wire precisely when a fracture occurs.

Consider just the wire-drawing process. Certain parameters relate to the incoming material, certain parameters to the outgoing material, and certain parameters are fixed for the entire spool. Control parameters change with every meter of wire. One can use the information gathered on fractures to make changes in the microprocessor controller during the process and affect key process variables in the outcome. The algorithm for control can be changed, the results of experiments observed, and new changes introduced. The decision maker can use this information in both day-to-day production decisions and in the development of algorithms to create new process capability over the long term.

The systemwide structure of IMC permits a plant to be organized into virtual cells for problem solving. These cells can be either horizontal (i.e., logical groupings of different machines) or vertical (i.e., logical groupings of machines of the same kind). The actual configuration will be related to the point of view (i.e., product, process, or manufacturing) and the problem being solved rather than to the location of machines.

Consider the horizontal cell in Figure 2-4, in which the output of several dry wire-drawing machines is passed through furnaces to a chemical bath. If suboptimal procedures in the furnaces can be detected, that information can be fed forward to make compensating adjustments to parameters associated with the bath. Or consider the vertical cell in Figure 2-5, which shows a number of wet wire-drawing machines that produce strands that are wound into cable. Current practice with wet wire-drawing machines is to make the output as consistent as possible and wind strands together randomly. With IMC, it becomes possible to know the precise diameter of each strand from each machine and thus to wind strands selectively so as to produce cable of just the right gauge. At the dies themselves, process control today relies on heuristics rather than understanding. With sensors, information can be fed back within a single die, thus gaining another level of intelligence.

The IMC system described above is event-based, statistical in nature, and attention-focusing. It allows construction and subsequent refinement of models of production processes. These char-

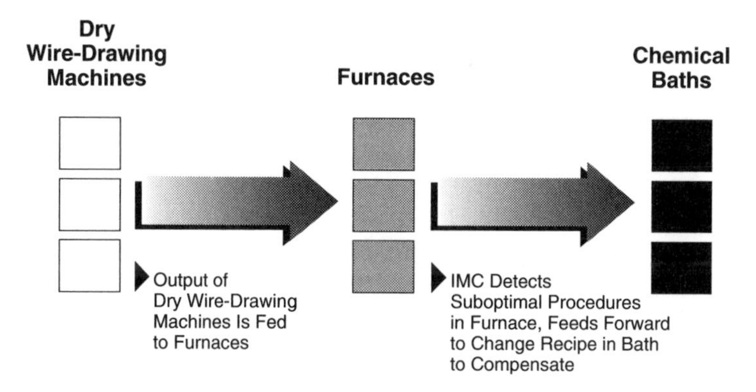

FIGURE 2-4 Horizontal logical cell employed to exert interprocess control between different operations.

acteristics permit distinctions to be made between systematic and random events, detection of novel events and patterns of events, and evaluation of features of interest. A complete picture of events can be captured and compared to expectations and alternative procedures. Economic values can be assigned to these procedures and relevant variations and experiments can be suggested.

Within eight months of developing an IMC system to control wire drawing on one line, fractures on that line were reduced fourfold. In addition, the firm expects to realize a tenfold increase in productivity. Because the architecture being used is general, many different implementations are possible.

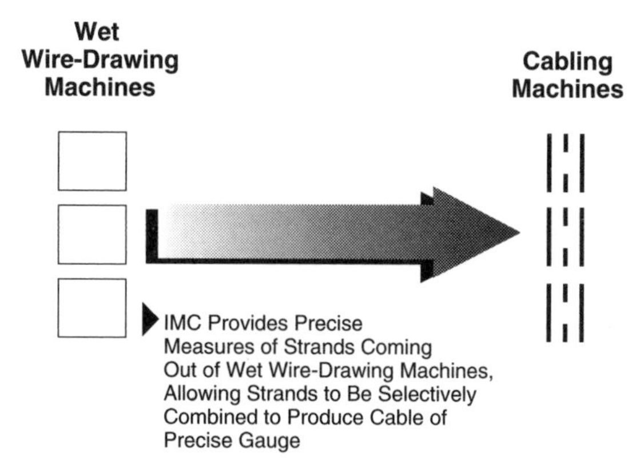

FIGURE 2-5 Vertical logical cell used to enhance control across machines of the same type.

IMC in the Chemicals Industry—Five Years from Today[7]

This case examines the future organization of a chemical plant that reflects the implementation of a host of existing or developing advanced process control technologies such as:

- three-dimensional (3-D) computer-aided design (CAD) imaging with walk-through and a computer interface (X-window);
- object-oriented modeling;
- multiple, integrated views of project design data, including an integrated dynamic process model;
- a standard interface that allows transfer of dynamic data between the plant control system and the CAD system;
- expert systems that provide embedded explanations of design concepts and current states of control systems;
- fiber optic sensing with embedded diagnostics;
- chemometrics with neural computers providing on-line composition;
- global access to process data; and
- natural language translators.

These technologies provide the operator's only view of processes that take place entirely within pipes and tanks—the notion of operating from behind a wall. Hence, the case plays out entirely in the control room.

In this hypothetical chemical plant, operators no longer spend 15 or 20 minutes catching up at shift change. Everything they need to know is now in the control room log, which is integrated into the business and maintenance systems. An incoming operator can tell, for example, that the product the plant switched to last shift, Betafon-134, is right on target in terms of the forecast for orders, and that the Zexene column has fouled and will have to be cleaned, a condition that will warrant analysis during the operator's shift.

Interaction with the operating system has become much easier. A large, flat wall display has replaced numerous cathode ray tubes for systems with different user interfaces. The new display is used by a distributed control system and several host computer systems. It has a single user interface: a glove that the operator can point at any part of the display and a headset for giving instructions to the integrated control system.

The display uses 3-D CAD imaging instead of menus and faceplates and gives the operator a videolike view of the process.

To examine the fouled Zexene column, for example, the operator voices a command into the headset mike and an exploded view of the Zexene column appears in an overlay. After voicing another command that causes the column to flash in blue, confirming that it is the right one, the operator asks the expert system for current data. Pertinent information overlays the image.

Suddenly, the operator is alerted to a more immediate problem—the Topper distillation column is flashing yellow, with the rest of the process running normally. A high-level diagnostic indicates that the column's control system is no longer in the normal state. This multivariable controller, which usually regulates controls at both ends of the column to on-aim control, has diagnosed the failure of its tails analyzer. The sensor's on-line diagnostics indicate that the fiber-optic probe has failed, causing the control system to rely on its on-line model to predict composition. The system has already automatically sent a priority electronic mail message to the analyzer specialist. The operator asks the system to predict the cost of the analyzer outage and is told that the model-based control system is compensating well but significant degradation in quality is likely. The operator asks the system to analyze the cost of switching to another product while running under degraded control. The results show that a switch to Gammafon-39 will allow the model to predict the tails composition much more precisely. A check of the integrated production planning and scheduling system indicates that the switch will not disrupt customer shipments, although it will increase plant costs somewhat. The operator uses the special glove to select Gammafon-39 from the product menu, voices the command to make the change, and then dispatches an electronic mail message to the team leader.

Frequent problems attended the old analyzer's sample system. Because the system could not explain its changes, operators often went manual when it did something they did not understand. Often the plant ran for days with bad composition data without the operator being aware of the failure. The fiber-optic probes of the new chemometric sensor look at the process stream directly, eliminating the need for a sample system. The operator does not need to understand the calculations in the neural computer to be confident that the system will accurately predict product composition and diagnose any problems.

Another multivariable controller indicates a reduction in monomer recycle to the Step A reactor. The on-line expert system informs the operator that the control system has detected a subtle change in catalyst activity, necessitating a reduction in the recycle rate for a short time until the catalyst can recover.

The operator is interrupted by the plant support engineer, who has dropped in to review some preliminary information on a new project aimed at increasing plant capacity. On a corner of the wall display, the engineer brings up a 3-D CAD image of the upgrade design, which is a composite of a conceptual design prepared by the corporate engineering design division and some detail design from the local regional engineering office, both sent to the plant electronically. Additions to the process are highlighted in green, modifications in yellow, and the existing process in blue. The engineer pulls up a schematic of a new reactor on the screen, notes the calculated residence times, and sees the plots from a simulation done by a consultant. The operator observes that the feed to this reactor from the existing process is very erratic. The engineer, wondering what effect this would have on reactor effluent composition, obtains from the plant control system flow data from that part of the real process for the past two months and uses the data to drive a simulation embedded in the conceptual design. The results suggest a problem, and operator and engineer decide to send the conceptual design, together with the plant data, back to corporate engineering design.

The operator, gesturing with the glove and voicing commands to recall the Zexene column display, reviews various formats that suggest that increased plugging in the column may be related to reduced catalyst activity in the Step A reactor. The operator runs analyses of last year's plant data and data for a similar plant in Japan and finding a correlation in the data from the Japanese plant, sends an electronic mail message to the plant engineer.

The phone rings and the operator is soon involved in a conversation that is a mixture of English and Spanish. A Spanish swimwear manufacturer needs a new Betafon product for a new swimwear line that adheres to some very tight specifications. The English-Spanish dialogue was a result of operator and customer mastering one another's language using a computer-based translator that provides interpretations of the conversation in the language of choice. Unable to provide an immediate solution, the operator calls an engineer at the Japanese plant and then calls the customer back. For several minutes, the operator (in the U.S.) and the engineer (in Japan) pan back and forth on the display screen, monitoring key variables throughout the process while the customer (in Spain) explains the problem. The customer sends the engineer a computer model—generated by a modeling package provided by the chemical company—describing the properties needed for the new line of swimwear and thanks the operator for making the connection with the Japanese engineer. In coming weeks, operator, engineer, and customer will become a tightly knit team as they work on the new Betafon product.

PRIORITIZED RESEARCH RECOMMENDATIONS

The panel believes that research in IMC must be directed at developing techniques for breaking down and refining knowledge as a foundation for building knowledge bases that are capable of adapting to change. Promoting synergy between people and machines is an essential part of this task. Research aimed at producing a world model for IMC should focus on high-level supervisory control that links both depth and breadth of knowledge. Research is also needed to develop data communication standards, sensor integration, and mechanisms for facilitating learning in an integrated environment. The necessary research must be jointly undertaken by industry and academe and must employ the factory as a laboratory, a theme shared with Chapter 3, Equipment Reliability and Maintenance.

In prioritizing its research recommendations, the panel concluded that productive areas of research in IMC lie at the cell level. In a world of dynamic product and process change, manufacturing must go beyond statistically controlling processes to building process capability.

To meet requirements for adapting intelligence to changing knowledge and organizational structures, knowledge bases must be developed that adapt to changing people, products, and external forces. Such development must be based on techniques for breaking down and refining knowledge. An information structure using these techniques, and operating in an uncertain, dynamic world cannot be built on machine intelligence alone; human–machine integration is essential. This synergy between people and machines must yield knowledge acquisition techniques that are capable of supporting rapid start-up.

A model that encompasses design, manufacturing, and management and can adapt to change also is needed, as is research on a variety of hybrid open-loop control systems.

The development of common standards for data communications, essential to the diffusion of IMC, implies creation of a specialized vocabulary for development and change. This approach to standardization must be technique-oriented and utilize standard physics, mathematics, business, and economics models.

Efforts at sensor integration should be guided by the need (1) to talk to the same world model as actuators, and (2) to integrate data, pattern recognition, and action models. Transparent algorithmic structures that facilitate ease of understanding and change and guide algorithmic development are very important for diffusion of the technology. Ancillary requirements include statistical control of process capability and an ability to reason from incomplete knowledge.

In addition, educational approaches must move away from training students and researchers, recognizing important work, and posing problems within specific disciplines. The laboratory—where theories are tested by experiments that control for potentially contaminating noise, the questions posed are narrow and well defined, and the results are unambiguous—is no longer sufficient as a locus of research. To make meaningful contributions today, research must reflect the union of engineering and manufacturing. It must recognize that the interfaces and interactions among processes have become as important as the processes themselves. Researchers must construct new methods of building knowledge and of unifying knowledge in different disciplines. The factory must become the laboratory because only in the factory can manufacturing be studied as a whole.

The context for research will demand close interaction between academe and industry. Research on systems that encompass entire factories cannot be done by academe alone and basic research in such areas would require a commitment of human resources beyond what most firms can afford. Interdisciplinary research, therefore, must be directed at this task.

This joint approach presents problems on both fronts. In academe, incentives for this kind of research are rare. Though the scope of such research is broad, because it is field-based and development is tied to a particular site with its associated idiosyncracies, it is not readily accepted and does not further an academic career. Furthermore, the track record for interdisciplinary research involving both engineering and management is not very encouraging.

Similar problems exist in industry. The factory is not viewed as a laboratory, and management of knowledge acquisition as an important, continuing activity is not part of the factory culture. Most U.S. factories operate incrementally, realizing only marginal improvements from the status quo. Their operation is not based on any vision of the future, let alone the vision presented here. While a firm may welcome a specific solution to a pressing problem, it is not likely to be interested in solving general problems of a basic nature in its factories (e.g., factory cost-accounting systems.[8] Mechanisms are needed that encourage cooperation between academe and industry and can accommodate conflicting goals, such as scholarly publication versus proprietary considerations and the need to experiment with real processes in real factories. In addition, management must be made aware that the rich stream of information guaranteed as a by-product of running its factories to produce products could be used to solve yield problems. The promise of IMC is to make the factory a more effective laboratory, capable of realizing both quantum and incremental improvements.

These barriers notwithstanding, industry perceives a need for training a new kind of engineer and for developing methods for managing learning and knowledge. Academe, in search of relevant initiatives, is preparing to meet the challenges of a broader playing field. Building intelligent systems that exploit person–machine synergies for learning in IMC is a significant challenge indeed.

In summary, research in IMC should aim at:

- developing technique-oriented communication standards to facilitate the diffusion of IMC;
- refining sensor technology in the areas of data integration, pattern recognition, and actionable models;
- building knowledge bases of design, manufacturing, and management intelligence that can adapt to changing knowledge and organizational structures;
- creating a dynamic world model of manufacturing;
- identifying ways to utilize the human–machine interface to facilitate learning in an integrated environment; and
- redefining its methods to accommodate holistic research in a production environment—the factory as laboratory.

MECHANISMS FOR DIFFUSION AND IMPLEMENTATION

Some of the larger Fortune 100 companies, in industries most threatened by foreign competition or in process industries that already use a high degree of closed-loop feedback control, may develop and build IMC systems independently. Such developments, however, will constitute isolated instances of technological proficiency that will diffuse only very slowly to the rest of the manufacturing world. (Witness the slow diffusion of robotics technology to the small-business community.) The panel believes that to move IMC to the larger manufacturing community, including enterprises with fewer than 100 people, rapid diffusion must be made an explicit focus of research on the architecture and development of the technology. Rapid diffusion can only occur if the building blocks become commodity-like elements that easily can be incorporated into the manufacturing system. In addition, the manufacturing community must have its eyes opened to the urgency of the competitive global challenge and IMC's vital role in addressing issues that may not be amenable to technical solutions.

The lack of diffusion of a similar technology, mechatronics, the integration of machine controls with electronics, suggests that early developers paid little or no attention to this requirement. Machine builders, using mechatronics, built sophisticated systems for large companies with specialized needs rather than general purpose systems for the larger body of small users, for whom the infrastructure for effective and easy use of a technology is as important as the technology itself. IMC systems, by their very nature, integrate all of the process steps in a manufacturing plant and require deep knowledge of each of those steps, so that any program that does not simultaneously build the infrastructure for the development of intelligent systems along with generic software for system integration will fail.

It is very important that an effort be made to diffuse knowledge about IMC rapidly. A cadre of good researchers and research sites must be built that will promote effective collaboration between industry and academe. Missionary work is needed in building an infrastructure for diffusion and in emphasizing the importance of the problem.

Even more important is the supply and building of talent and development of the necessary incentives for industry and academe. Scholarly publication of interdisciplinary research and effective peer review of such work is crucial to creating these incentives as is research funding and matching funding from business. Still another need is the creation of incentives to promote the development of educational and training materials that will enable instructors in universities, community colleges, and technical institutes to further diffuse the requisite knowledge to appropriate users.

Consider the application of the personal computer to manufacturing problems. In less than 10 years, the personal computer has profoundly influenced the way many manufacturing-related processes are performed. The rapid acceptance of this technology is due not only to dramatic price/performance reductions, but also, and more importantly, to the ease with which the average person can use the personal computer to solve problems and increase his or her productivity. The personal computer will surely become one of the building blocks of the future IMC system; it already has, to some extent. Other such standardized solutions must be found.

The panel recognizes that real-world management cannot abruptly move into the world of IMC—that move will have to be economically justified and made incrementally. IMC cannot simply be dropped into place in the manufacturing world. Its adoption is an evolutionary process that will have to be engineered to suit different environments. In a world economy, this is a vital process.

NOTES

1. Bohn, R. and R. Jaikumar. 1989. The Dynamic Approach: An Alternative Paradigm for Operations Management. Harvard Business School Working Paper No. 88011. Boston, Mass.: Revised August 1989.

2. There is a special reason for concentrating on the statistical aspects when introducing a program of better quality at lower cost in a going operation. They are more tangible than other quality control aspects and can be presented in a more interesting and appealing manner. The preparation of a list of trouble spots converted to costs per unit period, and plotted as one would a curve of cumulative wealth, will point out the operations where X (average) and R (range) charts should first be applied. (Juran, J.M. Quality Control Handbook. 1962. 2d ed. McGraw-Hill Book Company, New York, N.Y.)

3. The project was abandoned due to lack of continuing funding. ITI worked with the NASA sponsored Center for Autonomous Man-controlled Robotic and Sensing Systems, located at the Environmental Research Institute of Michigan. For further information, contact Dr. Robert J. Bieringer, Manager, Sensors and Control Systems Engineering, Industrial Technology Institute, Ann Arbor, Michigan.

4. Jaikumar, R. and R. Bohn. 1986. The Development of Intelligent Systems for Industrial Use: A Conceptual Framework. Research on Technological Innovation, Management, and Policy. 3:169-211. Boston, Mass.: JAI Press, Inc.

5. Jaikumar, R. 1986. Postindustrial Manufacturing. Harvard Business Review (November-December) 69-76. Reprint No. 86606.

6. Clark, K., R. Henderson, and R. Jaikumar. 1989. A Perspective on Computer Integrated Manufacturing Tools. Harvard Business School Working Paper No. 88-048. Boston, Mass.: Revised January 1989.

7. This scenario was adapted from material provided by E.I. du Pont de Nemours & Company and is used with that company's permission.

8. Jaikumar, R. 1990. An architecture for a process control costing system. Chapter 7 in Measures for Manufacturing Excellence, R. S. Kaplan, ed. Boston, Mass.: Harvard Business School Press. 193-222.

3

Equipment Reliability and Maintenance

SUBSTANTIAL CAPITAL investments, in the form of facilities and equipment, are required for manufacturing almost all goods of economic significance. The productivity of these investments is a fundamental element of competition among companies and nations. Events that slow or interrupt the manufacturing process or degrade the product impair the competitiveness of a manufacturing enterprise.

The term *equipment reliability and maintenance* (ERM) encompasses not only equipment, such as machines, tools, and fixtures, but also the technical, operational, and management activities, ranging from equipment specifications to daily operation and maintenance, required to sustain the performance of manufacturing equipment throughout its useful life. This chapter addresses all causes of diminished or degraded output. The panel considers ERM to be a significant factor in the competitiveness of manufacturing firms, an assessment supported by the case studies in the section on present practice (pp. 57-63).

Historically, the evolution of ERM can be traced from breakdown maintenance and repair to preventive maintenance to predictive maintenance. Breakdown maintenance and repair is the after-the-fact restoration of failed equipment. Preventive maintenance is the systematic servicing of equipment to reduce the possibility of failure. Predictive maintenance, in use in U.S. industry for only four or five years, is usually understood to involve the use of computer software to detect conditions that might eventually lead to equipment failure. Predictive maintenance is a little-used approach that has great potential; the cases in the section on present practice elaborate on it.

A few basic definitions are used in this chapter. The *theoretical capacity* of a manufacturing process or machine is the output per unit of time of continuous operation at the maximum safe operating speed. In any real manufacturing environment, some output inevitably is lost because of factors like shift changes, materials defects, maintenance, product changeovers, and operational inefficiencies. Most process engineers employ a deterministic safety factor approach that reflects a reasonable level of unavoidable loss of output. They design processes for a theoretical or running capacity, which is calculated on the basis of a system's estimated efficiency.

In actual, day-to-day production, manufacturing engineers generally monitor two basic measures of equipment or process performance, *scheduled* and *actual.* Scheduled output is the output expected from an operation for a given allocation of time, material, and labor; it is usually based on a published output rate. Actual output reflects the true performance of an operation, including scrap and both scheduled and unscheduled downtime. These figures are often given on a yield basis (e.g., output per unit of time or material input).

These definitions are important when comparing data for different firms in an industry. A conservative specification of rated capacity, however, can yield misleading data for equipment performance.

IMPORTANCE

ERM affects drastically the three key elements of competitiveness—quality, cost, and product lead time. Well-maintained machines hold tolerances better, help to reduce scrap and rework, and raise part consistency and quality. By increasing uptime and yields of good parts, ERM can reduce capital requirements, thereby cutting total production costs. It also can shorten lead times by reducing downtime and the need for retooling.

The replacement and displacement of conventional electromechanical factory equipment by mechatronic equipment have given rise to a very different set of reliability and maintenance requirements. The recent rush to embrace computer-integrated manufacturing (CIM) has further increased the use of relatively unknown and untested technology. The factory is becoming a web of interdependent subsystems, interconnected by computer controllers that communicate horizontally across peer processes and vertically to supervisory controllers above or slaves below. Much of the controller software was written with the assumption that all equipment works properly when, in fact, complicated and unpredictable failure modes, unanticipated by the system and equipment designers, are becoming increasingly apparent. It is seldom possible to predict how a system will fail when something somewhere in the plant breaks down. Because causal relationships are frequently hidden, repair is often time-consuming, expensive, and tedious. A better job of debugging these systems via simulation, analysis, and rapid development needs to be done.

Software and sensors have the potential to enhance ERM and so mitigate the added complexity associated with these technologies (see Figure 3-1). The trend toward integrating mechatronic equipment into factorywide systems provides a framework for extensively exploring and exploiting this approach. The potential

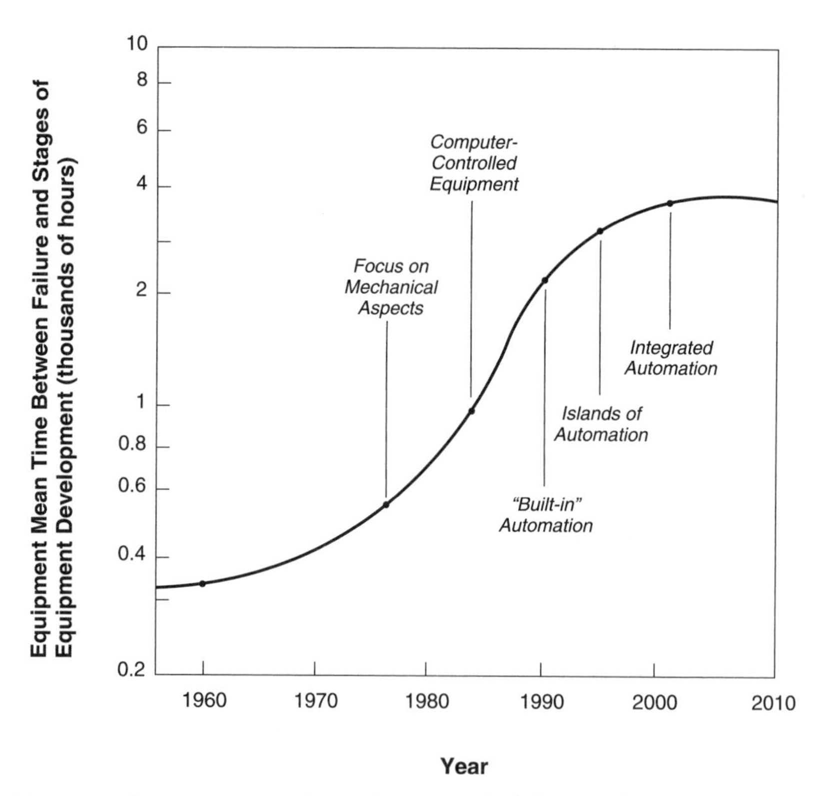

FIGURE 3-1 Improvements in equipment reliability with increasing computerization. Source: VLSI Research Inc.

impact of these integrated systems on capital investment alone is significant (forgetting impact on quality, cost, and lead time). If U.S. machine tool utilization, for example, were to go from 50 percent to 100 percent, investment could be cut in half.

PRESENT PRACTICE

In U.S. manufacturing, equipment availability is generally ensured by the existence of standby equipment, i.e., buying two machines when one will produce the desired throughput. Preventive maintenance, therefore, loses its urgency and breakdown maintenance becomes the order of the day. More importantly, the important maintenance issues do not surface.

The rapidly increasing complexity of factory equipment has led many U.S. manufacturers to look to suppliers for smarter equipment that will ease the need for a technically strong supporting infrastructure. This trend is producing a generation of catalog engineers who are adept at ordering equipment from a catalog but who lack the knowledge to evaluate equipment designs. Suppliers are responding by applying microcomputer technology to more phases of manufacturing operations and processing. The term *artificial intelligence* is appearing in sales literature that increasingly is offering catalog solutions as alternatives to technical proficiency.

This approach to maintenance is in marked contrast to that taken in other countries. In Japan, for instance, many firms are engaged in a 10-year program to upgrade predictive maintenance to the point where ERM becomes the responsibility entirely of operators. Attainment of set performance levels by equipment is celebrated with the sipping of *sake*. In firms whose management understands facility issues, including the implications of ERM efforts, just-in-time (JIT)[1] is less a measure of inventory than an indication that all processes are under control. Managers with this understanding can take a plant beyond the benefits of JIT to realize the potential benefits of ERM. The Japanese are succeeding at this because they have an integrated view of manufacturing as a system. Many U.S. manufacturing managers do not.

Japanese companies that have excelled in this area characteristically have the ability to develop and produce their own equipment. This is true even of smaller companies. Veteran engineers in one Japanese company with fewer than 50 employees, for example, developed all of the firm's equipment, tools, and dies. While this does not mean that U.S. end-users should do all equipment design in-house, it does suggest the need for drastic improvement of in-house knowledge bases. (See Chapter 6, Manufacturing Skills Improvement for more on this topic.)

U.S. industry is not without examples of emphasis on ERM. U.S. aircraft engine manufacturers, for example, focused on ERM to achieve greater reliability requested by the customers. Standards are published for maintenance and operation and for comparing performance of aircraft engines. Results of these performance comparisons are published in the trade press.

The ERM programs of several U.S. and Japanese manufacturers are outlined in the following paragraphs.

U.S./Japanese Automobile Part Stamping

Wholly owned and joint-venture manufacturing facilities established in North America by Japanese firms during the 1980s provide a basis for comparing U.S. and Japanese ERM practices in a variety of areas. One example is the stamping of automobile hoods, fenders, doors, and other major steel components.

The key technology in stamping is the transfer press, which incorporates a mechanism for moving parts between five or six dies. High acquisition and operating costs (stamping dies and presses constitute a substantial part of the capital investment in automobile manufacture) lead automakers to run multiple sets of dies on a given line. Die change time, which necessarily represents scheduled machine downtime, is a critical element of overall operating performance, and rapid, efficient die changing is a key basis of competition in stamping operations.

The U.S. facilities of Japanese automakers, using U.S. workers to operate three to five presses, often achieve at least twice the productivity of comparable U.S. facilities on a per-line or per-press basis. ERM practices contribute significantly to this achievement.

The ERM techniques of Japanese firms are basic practices that are carried out religiously. The Japanese stamping lines, for example, are specified to run at a lower, more reliable rate than equivalent U.S. lines. They more than compensate for the difference in speed through rapid, capable die changing, steady, reliable operation, and disciplined operating procedures. Dies are cleaned and lubricated routinely, and repairs are completed to original specifications before problems become major. Moreover, Japanese plants are characterized by a highly disciplined approach to basic housekeeping—floors and manufacturing equipment are kept spotlessly clean. The cumulative effect of these practices is reflected in the product—mass-produced automobile body parts that are comparable, if not superior, to those produced by manufacturers of low volume, luxury cars.

Near Defect-Free Pump Production

Between 1979 and 1982, the Japanese Nishio pump factory of Aishin Seiki reduced equipment breakdowns from 700 per month to effectively zero, while achieving an extraordinary level of quality—11 defects per 1 million pumps produced. Aishin Seiki's success reflects neither high technology nor large investments in new facilities. Rather, it is the result of a strategy for achieving zero defects and zero breakdowns through common-sense engineering, attention to basics, and hard work and team effort guided by a top-management vision. The strategy, which focuses on achieving economical life cycle cost for equipment and other physical assets, began in Japan in the early 1970s. It is a logical outgrowth of previous ERM philosophies stemming from the practice of preventive maintenance.

Aishin Seiki's ERM practices aim to maximize equipment effectiveness in terms of both quality and productivity by:

- establishing a total system of productive maintenance covering the entire life of equipment;
- involving all departments, including equipment planning, equipment usage, and maintenance;
- requiring universal participation by top management as well as shop floor personnel; and
- promoting productive maintenance through motivation management and autonomous, small group activity.

The goal of the strategy is to eliminate the six main downtime losses:

- losses caused by unexpected breakdowns;
- losses from setup time associated with periodic changes and adjustments of tools and dies;
- losses that result from idling, minor stoppages induced by sensors, and blockages of work in chutes;
- losses attributable to differences between actual and design speeds of equipment;
- process-related losses associated with defects and reworking; and
- yield losses that occur between start-up and steady production.

The strategy was implemented in four stages.

• In the preparatory stage, management (1) decides to employ the strategy, (2) mounts an educational campaign to reduce or eliminate resistance to it, (3) creates a hospitable organization and establishes basic policies and goals, and (4) formulates a master plan.

• In the introduction stage, the ERM program is kicked off.

• The implementation stage calls for (1) ensuring early successes and hands-on experience by focusing initial efforts on equipment that offers the most potential for improvement, (2) establishing an autonomous maintenance program for operators, (3) creating a system of planned maintenance that anticipates an increase in work load, (4) improving operating and maintenance skills through training, and (5) developing a prototype equipment management program.

• The final, or establishment, stage calls for (1) making the new ERM strategy the norm for conducting business and (2) providing a process for continuous improvement.

The Nishio factory is not an isolated showcase. Similar results are being achieved in old as well as new factories throughout Japan. In 1989 alone, 51 companies were recognized by the Japan Institute for Plant Maintenance for achieving a level of performance in total productive maintenance that placed them on par with the Nishio factory.

Diagnostic Monitoring of Steam Turbine Generators

With an aging complement of steam turbine generators and few new plants in the planning cycle, the U.S utilities industry is responding to increasing demand for electricity by adopting new practices for maintaining and improving the availability of existing equipment. Preventive maintenance practices in this industry evolved by trial and error. While they are generally effective, these practices have been considerably enhanced by state-of-the-art maintenance planning, management, and implementation technologies that incorporate information on equipment design, application, and operation.

One utility moved from manual plotting of stator performance data to a computer-based, continuous monitoring system on its 10-year-old turbines that were installed in the mid-1980s. The on-line system included hardware, software, and diagnostic engineering services. Initial emphasis was on detecting problems before they became serious enough to force outages. The system was later expanded to incorporate preoutage maintenance planning, the cost of generator outages, prevention of in-service failure, and the increasing percentage of older units in service.

Software problems during system installation and checkout and numerous false diagnoses attending a later upgrade incurred extensive debugging that involved both vendor and user. Still, the system has served the utility well. Problems have been diagnosed in time for corrective action, greatly minimizing damage to turbines. In one instance, the expert system embedded in the diagnostic system recommended that a generator in which it had detected a fault be kept in service until the scheduled outage. This correct diagnosis avoided both an unscheduled outage for repair and a probable forced outage later had the problem gone undetected during the scheduled outage.

ERM in the U.S. Semiconductor Industry

The cost of a typical state-of-the-art manufacturing facility for semiconductor memory chips has risen eightfold since 1980, from $25 million to $200 million. Over the next several product generations, the cost of such facilities is expected to reach between $500 million and $750 million (Figure 3-2). U.S. semiconductor industry capital investment of about $3.5 billion in 1988 strained the resources of even the largest companies.[2] Such investments warrant serious attention to ERM in this industry.

Since 1987, Intel Corporation has used ERM techniques to increase uptime and speed throughput for both the currently installed and the next generation of equipment. The company developed a two-part strategy that focused on reducing the operations and support costs of installed equipment while simultaneously influencing the design of a new generation of equipment using life cycle cost data from the factory floor. Partnerships with key equipment suppliers were critical to the success of both parts of the program. U.S. semiconductor manufacturers traditionally have not communicated with their original equipment suppliers, nor have they taken advantage of engineering and managing upgrades, modifications, maintenance, and spare parts issued internally. Japanese suppliers, in contrast, benefit from close customer technology exchange programs with leading Japanese semiconductor manufacturers. Intel and several of its key equipment suppliers have begun to turn the domestic situation around. Their technology exchange programs today include concurrent engineering of next-generation equipment for new product and process requirements and replacement of Intel factory equipment engineering and maintenance staff with supplier personnel.

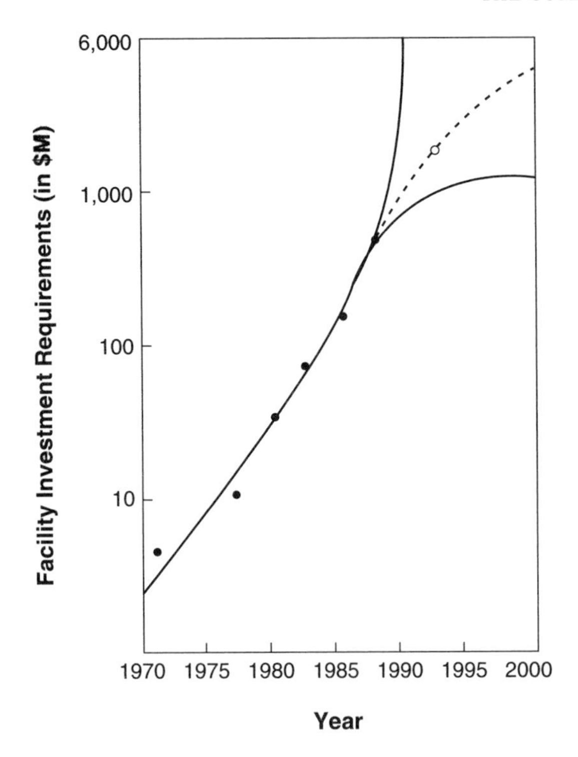

FIGURE 3-2 Expected trend in cost of state-of-the-art semiconductor manufacturing facilities. Source: VLSI Research Inc.

Key elements of the design/development phase include training and development to enable suppliers to perform highly specialized equipment analyses, such as failure modes and effects criticality analysis (FMECA), fault tree analysis, design for testability, design for fault tolerance/isolation/tracking, and design for remote diagnosis. Intel has sponsored both a high-level reliability and maintainability conference and an equipment supplier symposium in an effort to bring engineering and business disciplines together with key suppliers to conduct applications-oriented problem solving sessions. An equipment life-cycle training course was developed to train new suppliers, manufacturing engineers, and procurement teams. Procurement specifications have been upgraded to require that all key process equipment suppliers have a basic level of ERM expertise.

Results in trial fabrication plants for both wafer etch and implanter equipment show a substantial increase in average monthly uptime and increased predictability by reducing the range of monthly uptime (Figures 3-3 and 3-4). In photolithography, attempts to increase mean time between failures have yielded a 30-percent increase in availability and an approximately 30-percent decrease in area throughput time while achieving a 10-percent reduction in total fabrication throughput time (Figure 3-5). Mean time between failures for a wide variety of assembly equipment has increased in the range of 200 to 2300 percent, resulting in equipment uptime averages in excess of 90 percent of available hours (in a 168-hour week).

Benefits of Intel's ERM program have included reduced cost per unit at each process step, as factory operations and support costs associated with downtime or slower area throughput time have been reduced. Additional capacity can be made available in many cases by increasing equipment utilization rates rather than by adding units of equipment. This adds up to important savings because the base cost of many types of equipment exceeds $1 million per unit.

Indirect benefits often outweigh direct savings. For example, meeting process parameters more reliably via FMECA increases factory yields and reduces work-in-process inventory. Such continuous improvements will be required to keep direct cost per unit and factory overhead competitive in the global marketplace.

VISION

The ideal manufacturing environment of the future has no maintenance organization and every piece of equipment is expected to be available 100 percent of the time. The entire manufacturing organization, including executive management, recognizes the value of high equipment availability and the role of ERM in achieving it. Sensors monitor the condition and performance of equipment throughout the manufacturing system and feed back relevant data to system controllers that interact with technically sophisticated human operators who use the data to maximize throughput, effect timely maintenance, and recover promptly from equipment failure.

Equipment design employs new approaches that use broader data bases. Simultaneous design, or concurrent engineering, i.e., designing the production line or process at the same time as the product it is to produce, is the watchword. Designers have access to comprehensive data from the equipment itself, as well as a wealth of information about the needs and performance of the installed base of industrial users. Across each industry is a reli-

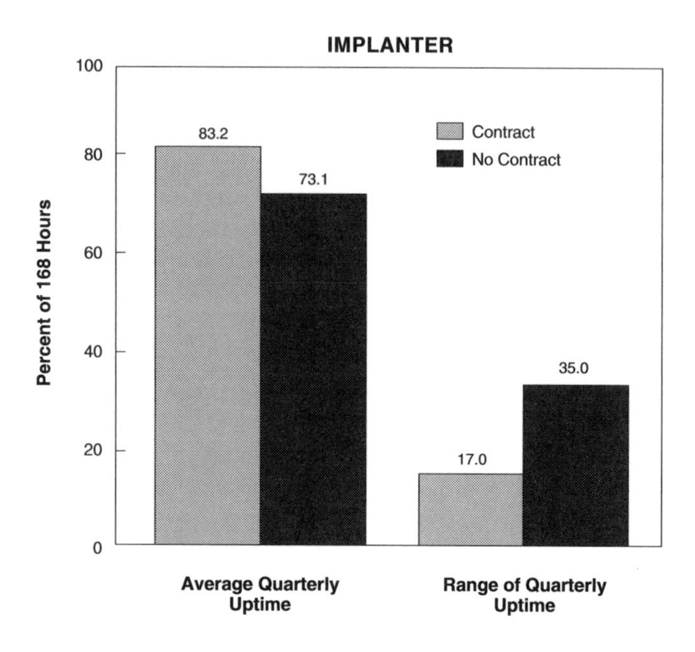

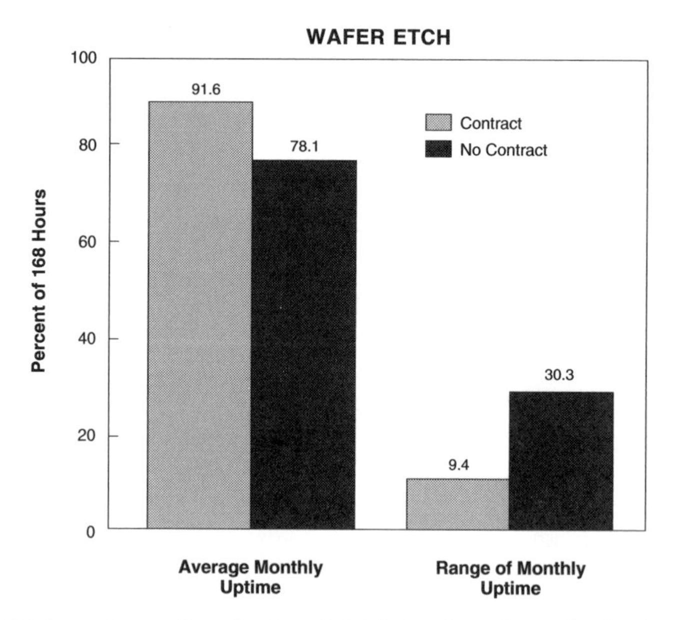

FIGURES 3-3 and 3-4 Results in trial fabrication plants for both implanter equipment (1987-8) and wafer etch equipment (1988), with and without service contract, show a substantial increase in average monthly uptime and increased predictability by reducing the range of monthly uptime. Source: Intel Corporation, 1989.

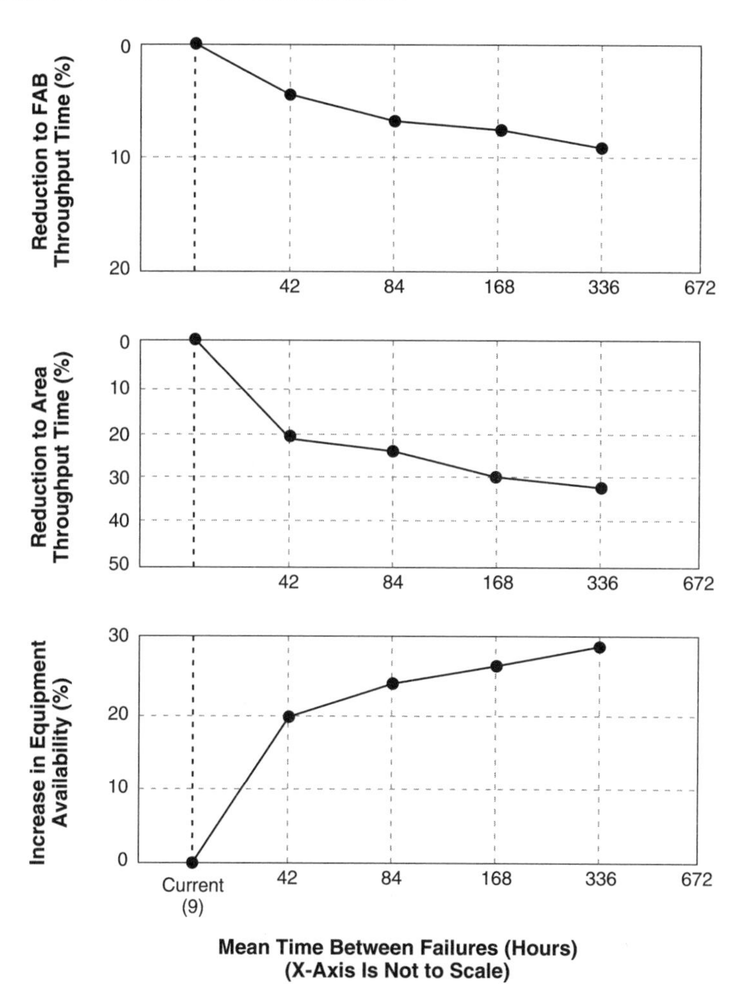

FIGURE 3-5 For photolithography, attempts to increase mean time between failures have yielded a 30% increase in availability and an approximately 30% decrease in area throughout, while achieving a 10% reduction in total fabrication throughput time. Source: Intel Corporation, 1989.

able body of knowledge, much like that available for engines in the aircraft industry today. Testability is a fundamental criterion for design, and an arsenal of reliable testing methods has given rise to equipment that facilitates experimentation.

U.S. manufacturing will want to jump to this futuristic plane from its present plateau. It cannot. This vision for ERM is rooted in the evolution of product and process design and intelligent manufacturing control. Present problems need to be resolved at a practical level, while high-technology approaches to ERM are developed in the next generation of equipment.

BARRIERS TO PROGRESS

Barriers to this vision of ERM are manifold. Perhaps the greatest is a general lack of awareness in the manufacturing organization, especially at the top, of the nature of ERM and its capacity for improving quality and productivity. The very terminology of ERM can impede its adoption and practice.

Even with greater awareness and understanding of ERM and sufficient good data to convince top management of its value, a serious deficit in the technical skills infrastructure required to implement an effective ERM program would still have to be overcome.

Lack of Awareness

In 1980, U.S. industry's utilization of capital and value-added time was low (Tables 3-1 and 3-2), its average machine utilization was less than 50 percent, and the country ranked lowest among the major industrialized nations in percentage of machines less than 10 years old (Table 3-3). Although more recent data are not available, the consensus is that the situation has changed little in the past decade.

Finance-driven managers who do not understand manufactur-

TABLE 3-1 Low Utilization of Capital

Equipment Category	Availability
Transfer Lines	50%-55%
NC Machine Tools	30%-70%
Conventional Machine Tools	30%-60%

TABLE 3-2 Low Value-Added Time

Work State	Fractional Time
Cutting	5%-6%
Auxiliary	4%-12%
Waiting	82%-91%

TABLE 3-3 Age of Machine Tools

Country	% Under 10 years	% Over 20 years
United States	31	34
Canada	41	37
West Germany	36	27
France	34	31
Italy	41	29
Japan	59	—
United Kingdom	38	24

SOURCE: Lawrence Livermore Laboratory. 1980. Technology of Machine Tools, Machine Tool Systems Management and Utilization. Berkeley, CA: University of California at Berkeley, October 1980.

ing operations and competitiveness can be misled by engineering and production managers who specify low rated capacities and scheduled yields. Capital authorization processes[3] provide only an appearance of effectively managing some of the crucial bases of competition, including ERM (on which production capacity is clearly dependent). Arguments that blame insufficient capacity on old machines that cannot be replaced because of high cost are suspect because the most cost-effective increase in U.S. manufacturing capacity may well be achievable through improved maintenance practices for existing equipment.

Manufacturing management has labored against prevailing views of manufacturing as an appendage to business, a necessary evil, bureaucratic and hence slow to respond, driven by internal pressures and short-term thinking, and not part of the strategic business plan. In general, manufacturing has been slow to recognize the value of the scientific base; it has placed inordinate value on activity (go fast to go slow) and subscribed to a trial and error approach to problem solving that fails to attack root causes. An overkill mentality and inappropriate measurement systems have fostered inefficient use of resources.

Many U.S. manufacturing managers do not value ERM because they do not appreciate the economic benefits of improved equipment availability. Making a case for increased spending on ERM to a management that perceives itself to be at a strong competitive disadvantage relies on demonstrating a benefit-to-cost relationship. But it is not yet known how to prove ERM's value using generally accepted accounting principles and short-term measurement systems that yield little information on life-cycle costs and benefits of equipment.[4]

Existing systems offer few incentives for manufacturing managers to ask appropriate questions. In fact, to the extent that they lead management to treat maintenance as an indirect cost that increases factory overhead, these systems may be putting the U.S. asset base at risk. Cost models associated with short-term measures, which rapidly drive future costs to zero in a present value analysis using a relatively high rate of return, lead managers to emphasize initial purchase price in equipment acquisition. In short, plant managers are given short-term cost-reduction targets that only can be achieved by deemphasizing maintenance staffs and equipment repair, the true costs of which will appear only over the long term.

Lack of Data

The hard data needed to make a convincing argument to management are hard to find. No available method of measuring system reliability (e.g., equipment effectiveness = availability × speed ratio × yield) has yet been widely accepted by U.S. manufacturers, and manually collected production downtime data are questionable, inasmuch as waiting time for repairs is seldom reported. As a result, it is difficult to make comparisons, share results and good practices among cooperating operations, and measure ERM performance trends in any one factory in a universally credible way.

Lacking standard measures of ERM and using reward systems that typically do not encourage operators or supervisors to record or try to maximize equipment performance, industry can give suppliers very little information about its equipment beyond installation. The suppliers, largely isolated, are in most cases not likely to commit resources to providing a capability that many U.S. manufacturers, because they have not yet understood its value, have not demanded.

Given customer preoccupation with initial cost and technical features, equipment is frequently designed to perform specific tasks within the constraints of an acquisition budget, with little regard for optimizing economic performance over its lifetime. Reliability—which may be the driver for a few key components or safety features, but is generally a secondary consideration for entire systems—is usually built in after the fact on the basis of experience and limited information. Suppliers are reluctant to project ERM performance, claiming lack of control over user application, adherence to recommended preventive maintenance schedules, and adequacy of repair. Firms that do value ERM, typically driven by powerful customer demands or regulatory influences, are forced overseas in their search for suppliers who are willing to quantify machine reliability and to work to meet ERM goals.

Lack of Skills

Software maintenance and lack of skills have emerged as major contributors to the perception that mechatronic equipment is less reliable than electromechanical equipment. Although companies technically unprepared to deal with these problems stand to suffer increased costs and reduced capacity, extreme pressure to minimize current-period costs continues to lead operations managers to sacrifice the hiring and training of maintenance personnel.

In contrast to design engineers who, because they are charged in most cases directly to a product, are viewed as contributors, manufacturing engineers and skilled workers are considered manufacturing overhead and, hence, as opportunities for cost reduction. In many companies, career paths for these employees are deemphasized, and responsibility for their training and development is specifically assigned to no one.

Lacking strength in manufacturing systems engineering, many plants rely on system integrators distinct from the equipment builders to configure their manufacturing systems. These outside suppliers of engineering services hesitate to experiment at the expense of their customers and so are unlikely to foster significant improvements.

This situation is reflected in the radically different perceptions of the performance of newly installed equipment in the United States and Japan. The U.S. view is that "the equipment is now performing at its best and over time performance will degrade." The Japanese view is that "the equipment is now performing at its worst and work needs to be done to improve its performance dramatically."

Emphasis on short-term profitability at the expense of preventive maintenance programs, personnel, and spare parts actually decreases long-term profitability by increasing the frequency of catastrophic events (i.e., breakdowns that demand emergency repair). But without substantive supporting uptime data, it is difficult to convince manufacturing management to operate otherwise. If the results of manufacturing technology research are to be effectively coupled to practice, a systematic, interdisciplinary attack—involving engineering, business, psychology, sociology, and finance, to name a few—must be mounted in order to change the status quo. To effectively promote ERM, a clear and strong initiative is needed to identify what needs to be changed, what the objective should be, and how to make the change, in order to have an immediate and significant impact with lasting effects.

RESEARCH NEEDS

In a highly competitive world market, a manufacturer that needs two machines to produce what another manufacturer, by assuring the reliability of its equipment, can produce with one, will not be cost competitive. The use of first-class equipment, operating at full capacity, is vital to competitiveness in the world market, a fact U.S. manufacturing managers must be made to believe and given the tools to act upon.

The panel has ranked research needs in ERM within two broad categories. One relates to people, the other to equipment (Table 3-4). Technology alone will not yield significant improvements in ERM—it is only an enabler. The problem needs to be viewed in its totality, with a balanced emphasis on human and technological issues and barriers.

Making progress with people-related issues will be much harder than solving technological problems. The task will require a major departure from the traditional lines of demarcation between the trades necessitating, in essence, a cultural change—probably the most difficult problem U.S. industry faces. While total abandonment of tradition, or complete cultural change, is neither possible nor desirable, U.S. manufacturers may need to import, adapt, and fine-tune techniques from other countries such as Japan.

U.S. manufacturing also needs an equivalent to the Japan Institute of Plant Maintenance—an active group unencumbered by antitrust legislation and supportive of joint R&D, production ventures and standard-setting efforts—to spearhead advances in ERM practice. Japanese efforts in this area have dramatically improved quality, lowered cost, and reduced plant breakdowns from 10 percent to as little as 1 percent of previous levels within three years. The charter of a U.S. plant maintenance institute might include: competitive assessment of ERM (United States versus other countries), research to improve ERM competitiveness and spread best practice, assistance to companies attempting to improve ERM, ERM training assistance, a forum for technology transfer, and recognition of companies that meet predefined ERM thresholds. Some specific needs in these areas are addressed below.

TABLE 3-4 ERM Research Needs

A. People-related needs:
  1. Executive management awareness of present ERM practice in the United States and abroad, and its importance for competitiveness.
  2. An understanding of enterprise optimization.
  3. A focus on using and managing technology.
  4. Academic emphasis on graduating career or professional engineers.
  5. Restoration of apprenticeship programs and manufacturing laboratories.
  6. General attention to the inadequacy of the basic level of talent.
  7. A new attitude that equipment should improve through use.

B. Equipment-related needs:
  1. Performance measurement
     a) Development of a standard methodology for measuring ERM.
     b) Development of industry-wide collections of ERM data.
     c) Development of a standard methodology for data analysis.
  2. Tools and techniques
     a) Design for reliability.
     b) Emphasis on sensor technologies and self-diagnosis.
     c) Development of systematic, disciplined methodologies that will enable manufacturers to communicate equipment requirements in clear operational terms.
     d) Development of guidelines and practices for designing the human–machine interface to enhance ERM.
     e) Development of standards for design and test, measurement and evaluation of ERM, and operator–equipment interfaces.
  3. Methodologies
     a) Analysis of concurrent engineering.
     b) Identification of operational changes to enhance ERM.
     c) Development of a strawman model for analyzing the impact of various methodologies on the manufacturing environment.
     d) Creation of a mechanism for providing equipment suppliers with the data they need to improve both uptime of installed equipment and mean-time-to-failure for new equipment.

People-Related Needs

Before anything can be done about equipment, U.S. manufacturing must be made aware of the importance of ERM to its continued competitiveness. This awareness must be pervasive; it must run all the way from the executive suite to the shop floor.

Effective management of ERM requires realistic awareness of the benefits of and necessary decisions related to ERM, the technical knowledge that will permit managers to make appropriate decisions at the conceptual level, and an understanding of the accounting issues involved. Management leadership is essential to solving manufacturing's problems because only management is positioned to have the overview essential for effective decision making and leading.

ERM issues will not be addressed unless management chooses to address them. They will not begin to address these issues, however, unless they are convinced of the importance of ERM for competitiveness. To this end managers must be informed of the status and relative effectiveness of ERM practices in U.S. and foreign industries. This would be helped greatly by a competitiveness assessment that reflects ERM differentials. Workshops sponsored by the National Research Council could serve to acquaint manufacturing executives with present practice in the United States and Japan. Because management needs both knowledge *and* data to know what to do and how to do it, a concurrent effort must be mounted to identify and supply the hard data without which even an enlightened management will be unable to identify the areas that need improvement.

If the vision set forth in this report is ever to be realized, the skills and capabilities that will be needed by managers at each level of the organization, and the agencies and institutions—universities, industries, companies, community colleges, etc.—best positioned to cultivate these skills, must be identified.

Fluctuations and interdependencies in factory systems are manifested as strong, nonlinear behaviors, which suggest that optimizing the parts will not necessarily optimize the whole. Too often, engineers work hard to improve one element of a system, only to discover that in doing so they have degraded the total system. An understanding of enterprise optimization is needed to guide the application of technical work in the plant.

Technology management is becoming an accepted discipline in a growing number of U.S. academic institutions. A complementary focus is needed on using technology. Quantifiable reasons must be found for employing technologists in the manufacturing sector of an enterprise. Well-understood methods of directing the work of these technologists are needed, as are mechanisms for implementing the results they achieve. Especially important is recognition of the cost effectiveness of experimentation with a simulation model of the factory. Managers and engineers should be aware of the benefits of using simulation methodology during process design. By allowing the ERM process to be modeled during system design, simulation fosters a total systems approach to design.

The gap between engineers and managers needs to be narrowed. Product and process engineers should be involved in manufacturing decision making, with financial analysts supplying guidance. Current emphasis should shift from graduating applied scientists and specialists to producing engineers and interdisciplinarians. Needed are more career or professional engineers from interdisciplinary programs that cut across engineering disciplines to also embrace elements of business, psychology, and sociology. These bridges outside the engineering school will help recast the industrial engineer of yesterday as the systems engineer of today.

In addition to a broader professional Master of Science degree, greater emphasis is needed on the production of master technicians. The unavailability of internship programs is a serious deficit in U.S. manufacturing.

Who is, and is not, walking the floors of today's plants is also of concern. Few U.S. manufacturers have programs that provide white collar staff with exposure and experience in the actual manufacturing process. Many of the managers who cannot improve their plants arrived in manufacturing with finance/MBA training and had never set foot in a factory. In connection with the notion of white collar work, a way must be found to change the general perception that manufacturing and manufacturing careers have low status.

Schools have followed business's lead in the past; when business deemphasized manufacturing in the 1960s, schools closed their manufacturing laboratories. There is no reason to expect that they will not respond similarly to a reemphasis on manufacturing. Some two-year colleges are already cooperating with manufacturers to teach technical skills to shop floor personnel. Manufacturing engineering and software training programs are badly needed. Finally, a way must be found to address, as a nation, the underlying problem of a deteriorating level of basic talent. Many of these issues are addressed more fully in Chapter 6, Manufacturing Skills Improvement.

Equipment-Related Needs

Equipment-related needs can be grouped under the categories of performance measurement (identifying what we need to know and how to measure it), tools and techniques (what we can do for and with the machines), and methodologies (generalized approaches to ERM).

Performance measurement

A standard methodology for measuring the performance of ERM is needed, as is a way to quantify the benefits of ERM for top management. Though generally accepted accounting principles are inadequate for manufacturing decision making, no substitutes have gained wide acceptance. Without a measurement system that can quantify relationships between engineering work and profit/asset benefits, assess the costs of equipment waiting (e.g., load, unload, operator, material) and downtime, and allocate maintenance costs to reflect reality, results will be anecdotal and perception of the basis for changing the incentive structure will remain shaky. A more enlightened approach to reducing equipment life-cycle costs would be to increase capital investment at the concept and design stages in order to improve equipment reliability, maintainability, and predictability and thereby reduce operations and support costs that inflate factory overhead (Figure 3-6).

To evaluate the cost of machine downtime, one must first know a machine's value per unit of time, which can be obtained through a variety of methods related to capacity utilization (see definition of theoretical capacity on p. 55). Needed is a production monitoring system capable of providing information about all machine shutdowns, classifying discrete events, and aggregating these events within categories. Given the opportunity cost of shutdown, such a system could calculate the benefit of reducing or eliminating each category of downtime. Further, a logical model of machine failure that can relate different kinds of failures, mean times between them, and their causes, is needed. Industrywide collections of such information will facilitate sharing of technical data—both ERM statistics and qualitative information about the relative effectiveness of different techniques—and help pinpoint areas where improvement is essential to global competitiveness.

Tools and techniques

Needed in equipment design is a design-for-reliability approach that takes into account all life-cycle costs (acquisition, installation, operation, maintenance, repair, and salvage). This approach should delineate basic principles that are demonstrably effective and universally applicable. ERM does not begin on the manufacturing

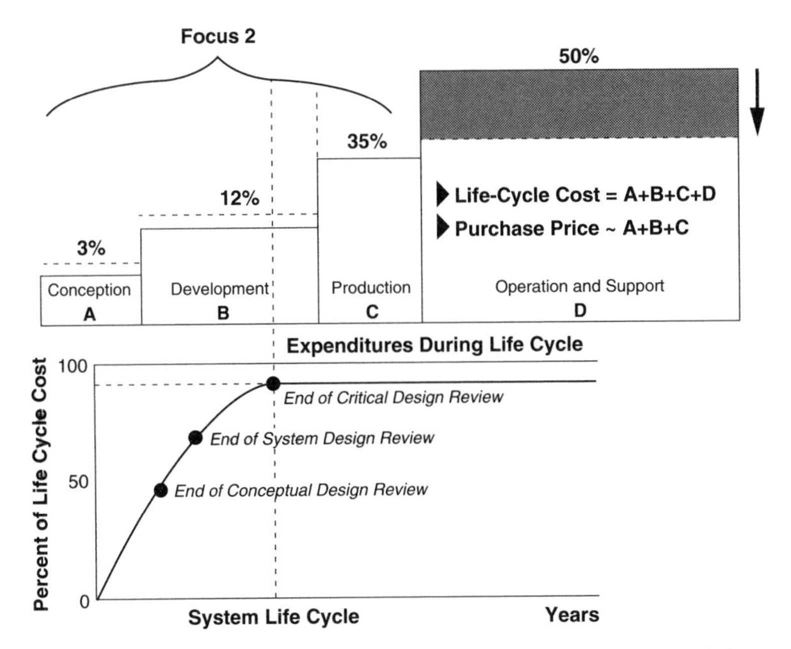

FIGURE 3-6 An enlightened approach to reducing equipment life cycle costs: Increase capital investment at the concept and design stages. Source: Intel Corporation, 1989.

floor; it begins in the office with the design of product, process, and equipment.

While judgment is required to ensure that manufacturing equipment is efficiently utilized, sensor technologies and self-diagnostics may be sufficient to ensure that it is reliably maintained. Research needs in this area are amplified in Chapter 2, Intelligent Manufacturing Control.

A clear link must be established between equipment requirements and reliability. Critical specifications must be defined and formulated to ensure reliability and identify pitfalls, stumbling blocks, and best practice. Systematic methodologies that will enable manufacturers to communicate equipment requirements in clear, operational terms need to be developed.

In the area of human–machine interface, research is needed to delineate the guidelines and practices needed for establishing this important junction.

Development of standards for design and test, for measuring and evaluating ERM, and for operator–equipment interfaces, would be a great facilitator in each of the areas discussed above.

Methodologies

Concurrent engineering warrants study as a means of building ERM considerations into equipment from the beginning. To this end, the training of finance and accounting people to guide and support the concurrent engineering process is critical.

Operational changes that will enhance ERM must be identified. One need is a checklist of procurement practices that will ensure that reliability, or lack of it, is recognized in specific equipment. In addition, practices that will promote effective implementation and continuous improvement of ERM must be developed.

A strawman model that would support analysis of the impacts of various methodologies on the manufacturing environment would be useful. Such a model should address matching of capabilities, design issues and rules, maintenance, installed base, culture, and organization. Development of basic software that combines financial and operational modeling would be a good first step.

Summary

Manufacturing is easily the most challenging and complex system in any organization. ERM must be recognized as an integral part of manufacturing that has a major impact on quality and productivity. Viewed as a discrete activity performed at different points in the manufacturing process, ERM will not achieve its full potential. It must be viewed as a system within the larger manufacturing system, a system that can be modeled in successful operation.

The research agenda should reflect the present perspective on ERM. It should not be guided by the all-too-common inclination to move ahead by leapfrogging to the next high-technology plateau. Rather, research should follow the model that takes the long-term view while working hard on short-term activities. The Japanese have done it by adhering to the basic principles of motivation, commitment and understanding, and training and discipline. Inquiries into ERM in the next generation of equipment should be based on practical investigations into improving it in the present generation.

To this end, research in the area of ERM should be directed at finding ways to:

- inculcate the attitude that equipment should improve through use;

- identify and analyze operational changes that might enhance ERM and inform equipment suppliers of these changes;
- communicate ERM requirements in clear, operational terms;
- support industrywide collection and analysis of standardized ERM data;
- foster an understanding of enterprise optimization; and
- make all of the foregoing accepted practice from the shop floor to the executive suite.

NOTES

1. Just-in-time (JIT) is a method of manufacturing by which parts and assemblies are made or delivered as needed, thereby greatly reducing inventory.

2. National Advisory Committee on Semiconductors. 1989. A Strategic Industry at Risk. Washington, D.C. November 1989.

3. In many companies, the capital approval process specifies different levels of authorization depending on the size of the request. Small investments (under $100,000, say) may need only the approval of the plant manager; expenditures in excess of several million dollars may require the board of directors' approval. This apparently sensible procedure, however, creates an incentive for managers to propose small projects that fall just below the cut-off point where higher level approval would be needed. Over time, a host of little investments, each of which delivers savings in labor, material, or overhead cost, can add up to a less-than-optimal pattern of material flow and to obsolete process technology. (excerpt from: Kaplan, R. S. 1986. Must CIM be justified by faith alone? Harvard Business Review 64(2): 87-95.)

4. Kaplan, R. S. 1989. Management accounting for advanced technological environments. Science 245 (August 25): 819-823.

4

Manufacturing of and with Advanced Engineered Materials

ADVANCED ENGINEERED MATERIALS (AEMs) are high value-added materials that invariably perform better than conventional materials, yielding products that, for example, are lighter, have broader service temperature ranges, are multifunctional, or have better life-cycle performance. The added value may result from either more expensive components (e.g., fibers) or a sophisticated or difficult processing sequence. Such processing must produce materials that satisfy both the geometric and property demands of the application without degrading them. Hence, control of processing details and near net shape technologies become increasingly important. Where standard product forms do not exist, or the market for these materials is potentially specialized and limited, considerable capital investment or flexible manufacturing schemes will be needed to establish reliable, reproducible production capability.

This chapter focuses on the unique processing and manufacturing challenges and opportunities of AEMs. It presents examples of classes of engineered materials and considers the state of the art of processing science and simulation as well as the goals of design for manufacturability and economic modeling and projections. Future needs and directions are presented from the focus of advanced materials, since their use can introduce barriers to optimization outside the normal scope of manufacturing science and engineering. Also considered are needed research in materials science and engineering, the need for expanded and revised educational programs and objectives, and methods for better integrating materials-specific issues in manufacturing.

Engineering applications that challenge the capabilities of conventional materials are often best addressed by a host of AEMs, such as:

- high-temperature structural materials (e.g., monolithic and composite intermetallics and ceramics),
- high modulus materials (e.g., liquid crystal polymers and supermodulus copper-nickel periodic layered structures),
- multilayered heterostructures (e.g., ternary/quaternary compound semiconductors, and silicon-germanium alloys),
- advanced coatings (e.g., diamond films and electronic packaging), and
- optical materials.

The scale of a material's structure may vary from macroscopic, as in fiber-reinforced composites or multiphase alloys, to molecular or atomic, as in multilayer semiconductors or copolymers. The inevitable presence of interfaces poses a unique set of challenges for both the producers of such materials and the manufacturers who process them. With fiber-reinforced composites, for example, understanding wetting and dispersion of fibers and concomitant interfacial reactions is critical to the design of manufacturing paradigms and prediction of system stability during use. In multilayer materials for electro-optic applications, whose properties stem directly from the ability to control and exploit processing parameters, controlling growth of pseudomorphic layers is not only integral to the manufacturing process, but is, in effect, the driver.

STATE OF THE ART

Issues of simulation and the scientific, engineering, and economic bottlenecks associated with AEMs, are examined in light of the following considerations:

- stability of microstructure and interfaces throughout processing, manufacturing, and use;
- lack of property and predictability data bases, and of appropriate institutional settings for disseminating such information;
- existence of niche markets and boutique materials;
- special processing requirements and manufacturing equipment, and transient and nonequilibrium processes;
- need for flexible and accurate process control;

• difficulty in identifying, measuring, and controlling critical parameters; and
• safety, environmental, economic, and educational issues in manufacturing.

These issues are examined first from the perspective of materials processing and manufacturing and then in light of the educational and training infrastructure that supplies processors and manufacturers.

Materials Processing and Manufacturing

To illustrate the critical need to understand and control processing better in the production and use of AEMs, some critical aspects of the state of the art are identified for two broad classes of composite AEMs, polymeric and metal-based.

Polymer-Based Composites

Problems in processing polymer-based advanced materials, both in synthesis and in conversion to useful articles, stem largely from the heterogeneous nature of the materials. Composite structures are a well-recognized aspect of materials such as continuous fiber-reinforced thermosetting polymers. They are also present, though not as obviously, in many other advanced polymers. For example, a toughened thermoplastic for automotive bumpers may consist of a miscible blend of ductile amorphous polymer with a crystalline polymer (to add solvent resistance), blended immiscibly with an elastomer for toughening. As another example, liquid crystalline polymers are highly ordered within domains, but the macroscopic orientation of the domains depends strongly on the material's processing history.

The properties of all of these composite and compositelike materials are functions of the microstructure—that is, of the sizes, shapes, and arrangements of the component phases or materials—and are determined by the nature of, and forces at, the interfaces. Designing processes that will result in the desired microstructural and interfacial characteristics and behavior presents a unique challenge. Traditional methods, such as winding a continuous filament in a desired pattern and impregnating it with a low-viscosity thermosetting prepolymer, or laying up prepregs, are suitable only for a limited class of applications. It would be useful to extend the above approach to thermoplastics to take advantage of the shorter fabrication cycles, increased toughness and repairability, and recyclability of scrap. However, the relatively high viscosity of thermoplastic matrix materials makes the wetting and impregnation of fiber bundles difficult, so that new manufacturing concepts are required to make these AEMs commercially viable.

Blend or alloy polymers, liquid crystalline polymers, and short-fiber reinforced thermoplastics are characterized by inability to control their microstructures during manufacture and processing into components. This situation is attributable in part to lack of understanding of the interrelationship between the materials' thermal and deformation history and their microstructure. Increased understanding in this area is essential for the invention of new processes and control mechanisms.

The sensitivity of specific properties to a material's microstructure also makes it difficult to predict resultant properties, even when the properties of the components are well known. Although this problem is most severe for ultimate properties, such as strength and elongation to failure, and for endurance properties, such as creep and fatigue, even the small strain properties, such as elastic moduli, are difficult to predict, particularly for design purposes.

Metal-Matrix Composites

Forming parts from metal-matrix composites (MMCs) customarily involves adapting methods that originally were developed for monolithic materials. Powder metallurgy and liquid/slurry casting, the two principal adaptations, are at best compromises that reflect constraints and options that include required performance, relevant properties, ease of process adaptation, and cost.

MMCs involve engineered macroscopic and microscopic arrays of two or more phases or materials that may not be in thermodynamic equilibrium, which often leads to interfacial instabilities. The result is that interfacial structural and chemical changes can occur rapidly in critical temperature-time conditions. In addition, neither property requirements nor processing methods are fully understood in relation to the detailed macroscopic and microscopic configuration of the multicomponent array and with the interfaces between the components in the array.

These considerations translate into several major concerns that must be resolved if MMCs are to become a major option for a broad range of manufactured products.

1. MMCs achieve unique properties through systematic combinations of different constituents with significantly greater processing complexity than is usually required for monolithic materials. The main concerns are structural (fiber-matrix) variables, property requirements, and processing options, including variables such as fiber composition, structure, relative dimensions and orientation in the matrix, nature and stability of bonding at the interfaces, and matrix mechanical and physical properties. These structure, property, and processing options form a three-dimensional matrix through which well-defined paths need to be identified to optimize properties and performance while controlling processing and structural complexity. For example, careful mapping may reveal that two or more presently used combinations are redundant, in which case a choice may be made on the basis of cost and manufacturability alone.

2. A fuller understanding is needed of how processing options upgrade or degrade the performance of MMCs. The properties of an MMC are optimized when a spatially uniform dispersion of oriented and undamaged fibers or whiskers is in intimate, bonding contact with a structurally and compositionally uniform metallic matrix having high (near theoretical) density. Such optimization is exceedingly difficult to achieve without favorable bonding characteristics between a dispersant and a matrix that have been thoroughly mixed. Good fiber/whisker wettability is an essential characteristic for an MMC processed by foundry methods.

3. Metal matrix-ceramic fiber combinations, whose potential for widespread use in manufactured products is promising, are currently optimized for in-service properties, but not for processability. Intrinsically poor fiber-matrix bonding/wettability is made acceptable only through complex, difficult-to-control processing steps, such as precoating fibers and/or alloying and premelting the matrix to improve its wetting characteristics. Both of these steps, which have been developed empirically, can degrade the physical and mechanical properties of the MMC, especially if excessive chemical interpenetration and structural alteration occur at the fiber-matrix interface.

Educational Infrastructure

Over the past several decades, a serious erosion of attention to materials processing in university materials engineering programs has been allowed to occur. Of the many factors that have contributed to this trend, two are relevant to the present discussion.

1. Growing emphasis in the literature on basic structure–property relationships has led to an increasing classroom and laboratory focus in this area.

2. Teaching of materials processing continues to focus primarily on traditional, mainly metallurgical practices. Processing of most of today's AEMs (metals, ceramics, polymers, electronic materials) relies on methods that are foreign to traditional metallurgical practice, and, therefore, are unfamiliar to faculty who do not have contact with industry.

In short, materials processing has lost ground in materials engineering education in terms of emphasis and technical timeliness. The timing of this lag is unfortunate because structure, properties, and processing are inseparably melded in AEMs, and the required processing is rarely a simple extension or adaptation of traditional processing methods.

VISION

The current trend is to look beyond the well-defined categories of materials such as metals, polymers, and ceramics. The materials of the future will be composites of all of these. Ceramic fibers are strengthening metal matrices; polymers, strengthened with woven fabrics, are providing radically new materials with improved strength and toughness and lower density. These techniques continually will be exploited and expanded.

Engineering is moving increasingly from simple use of those materials' properties that can be found in a handbook to the customization of materials' properties for precise uses. This greater differentiation in composition or form from traditional practice can lead to more value added during fabrication. This often leads to composite structures. For example, the most sophisticated packages of very large scale integration (VLSI) chips are made from interpenetrating layers of metals, ceramics, and polymers, requiring advanced processing to create micro- and macrostructures designed to yield desired performance characteristics.

The key to success in this arena has been continuing development of composite materials and concomitant control of microstructures. One result has been that knowledge of materials and components has greatly expanded from the late 1950s to the present. An extensive knowledge base has been built in metals solidification, for example, allowing metal micro- and macrostructures to be correlated with both the solidification rates of alloy systems and the particular equilibrium thermodynamic relationships of a system and its various kinetic coefficients. Although this knowledge is directly applicable to the solidification rate behavior of molten ceramic alloys, whose equilibrium thermodynamic and kinetic relationships have parameters and functional forms similar to those of the metal alloys, extension to more complex systems is not yet possible. This lack of a more sophisticated data base clearly inhibits exploitation of AEMs.

Versatile new processing methods for electronic materials—including plasma-enhanced etching and deposition, lasers and other high-energy beams, and high-energy radiation for photolithography—greatly increase packing density in integrated circuits and offer new possibilities for high-speed and optoelectronic devices. Submicron features made by electron beam, X-ray, and synchrotron radiation will permit the development of integrated circuits (ICs) that have many millions of transistors and capacitors properly interconnected on a silicon chip. The use of plasma-enhanced processes, flash lamps for annealing, high-pressure oxidation, ion implantation, and other advanced processing methods permits fabrication of semiconductor devices at much lower temperatures than is possible by more traditional means, leading to devices with smaller geometries and fewer defects. New crystal growth techniques, such as molecular beam epitaxy and organometallic chemical vapor deposition, which not only allow growth of multilayer heterostructures of compound semiconductors, but also the combination of silicon with III-V compounds (e.g., gallium arsenide and indium phosphide), are extremely promising for a variety of high-speed and optoelectronic devices and systems.

CHALLENGES

The progress in materials processing and development of new, complex systems, though impressive, has resulted largely from empirical research and nongeneric development, often separated by distance and philosophy from the production environment. Clearly, this must change.

An emphasis on manufacturing technologies will be critical in the early stages of any new materials development. Because processing affects every aspect of the properties of AEMs, processing methods should be integrated into design and development from the beginning. At the same time, to ensure that manufacturing steps do not degrade properties of AEMs, sensitivity to materials' properties should be explicitly integrated into design paradigms. Planning manufacturing steps so that final processing occurs during shaping and assembly, for example, might reduce production costs. Indeed, the entire process, from understanding the applications and technical needs of the customer through design and manufacture, including timely distribution, must be integrated into a well-understood and coherent procedure. The challenge today is to achieve excellence in each of the activities involved in the design, development, production, and distribution of AEM-based products to customers. Excellent functional capabilities must be integrated with other capabilities in the manufacturing organization to achieve superior overall performance. The critical functions in an AEM company include:

- acquisition and implementation of science and technology for AEM products and deep processes;
- complete understanding of customer needs and, more broadly, of emerging markets;
- superior production capabilities, including product and process design and implementation; and
- effective organizations and systems for sales, service, and distribution.

Process Simulation and Modeling

Because the AEMs of the future will offer a broader range of microstructural and process options, process simulations will be needed to help select materials development and manufacturing pathways. Materials engineers must be able to explore a greater range of options and predict the linkages between processing conditions and material microstructure, and between that microstructure and the behavior of the material. Computer simulation offers an opportunity to assess competing development approaches while reducing the delay and cost associated with prototype construction and experimentation.

AEMs will be required to adhere even more closely to their performance thresholds, which will require tighter quality control. Production of such materials will rely on improved process control that is founded upon a deep understanding of the controlling parameters. Process simulation is one route to this understanding, with the goal of developing intelligent processing of materials (IPM) systems.

Process simulation can facilitate rapid prediction of materials characteristics (e.g., shape, microstructure, and residual stresses) as a function of processing parameters, which leads naturally to appropriate control strategies. A successful simulator must embody the physics of the process in analytical, numerical, and heuristic models, using, where necessary, approximate solutions and precalculated results, and learning to increase system responsiveness. The simulator must run fast enough to hold the attention of the user through "what if" analyses of process design and operational options. The models should be patterned after computer models that simulate aircraft flight and operation; they must be fast enough to give the user the feeling of controlling the process, and user interfaces should rely heavily on rapid, real-time graphics and permit, as required, system reconfiguration and analysis of process output data.

The processes used to produce AEMs involve a broad spectrum of physical phenomena, which must be incorporated into the simulator models. Available models usually focus on a limited subset of individual phenomena, which themselves may entail considerable approximation and computation. Because most real processes involve many phenomena operating simultaneously and synergistically, the challenge is to develop an integrated set of models of the materials system that operate fast enough for the proper functioning of a simulator. One way to increase simulation speed is to establish a hybrid modeling approach that combines precalculation with enhanced numerical models founded upon reduced-order models.

Process simulation for materials and process development for IPM implementation requires models that predict the evolution of microstructure. Prediction of the influence of thermal and fluid flow and deformation processing on residual stresses and microstructure is especially difficult because of the relative lack of sophistication of microstructural modeling and the difficulty of symbolically describing microstructural features.

Ideally, the process simulator should be usable in several modes to enhance material quality and speed the move from materials development to production implementation. It should be able to deduce directly the combined effect of materials design and process planning on product characteristics and the effects of these characteristics on properties. It also should be able to determine which process parameters and intermediate microstructural characteristics exert the greatest impact on materials properties and product quality, thus facilitating identification and ranking of sensor needs and formulation of intelligent control procedures. Finally, the simulator should aid in diagnosing on-line production difficulties and replanning to correct deviations from the intended process trajectory.

Knowledge-Based Systems

Superior AEMs are expected to result from the use of expert systems technology to integrate systematically processing knowledge and materials science. The success of the reasoning tools needed to integrate symbolic knowledge of materials design, process planning, and IPM will depend on the understanding of the underlying materials science and the logic of the problem-solving approach employed.

To date, only modest success has been achieved in applying knowledge-based systems technology to materials systems. Planning approaches based on artificial intelligence (AI) are being used to aid in the design of conventional alloys, as demonstrated by the Aluminum Alloy Design Inventor (ALADIN) system,[1] jointly developed by Carnegie Mellon University and Alcoa Research Laboratories. Another achievement has been the use of expert systems concepts to incorporate symbolic process operator knowledge into the formulation of process control. Each of the above has demonstrated some success in either production or control, but not in more clearly delineating the formidable problems that remain.

More powerful expert systems are clearly needed to represent more expressively materials and process engineering knowledge and to facilitate operation of intelligent control systems. A major challenge is to build intelligent controllers that are capable of translating materials and process understanding and reasoning approaches into planning and control formulations that can be executed by computer programs.

Another challenge is related to the communication and display of scientific information. Materials scientists often use geometrical representations (e.g., graphs or pictures) to express phase relationships and microstructural and fractographic descriptions. Words and numbers cannot adequately express geometric information, and because AEMs incorporate an extremely wide spectrum of microstructural features that span several hierarchical levels, no single photograph or diagram can encompass them all.

Materials scientists can infer the succession of phases, as well as phase chemistry and volume fraction, by tracing the multidimensional pathway through phase and transformation diagrams, the most commonly used means of relating phase constituency and temperature dependence for complex alloys. Reasoning is performed by coupling the graphical thermodynamic and kinetic representation with other microstructural and experiential knowledge, with the aim of solving these problems and others that are sure to emerge. Developing paradigms and software to facilitate efficient component manufacture has the ultimate goal of producing high quality materials that are free from damaging flaws, residual stresses, and microstructural abnormalities.

IPM is a powerful processing concept that promises to integrate fully process knowledge, models, sensors, and control technology. The successful IPM control system will use an AI-based controller to establish processing plans and coordinate operation of a conventional dynamic control system in order to provide consistent and logical processing plans and decision making throughout the cycles of materials development and production implementation.

Development of such intelligent control relies on advances in understanding of materials processing, AI-based planning, and control technology. Determination of materials' microstructures, residual stress states, and processing pathways and identification of related reasoning strategies are all within the province of materials scientists and process engineers. These factors need to be reduced to a coherent generic format before an intelligent control design can be developed and the necessary software implemented.

Sensors

A critical part of IPM systems is sensors that provide highly detailed macroscopic information on dimensionality, position in space, shape, velocity and acceleration, global and local temperature, and compositional distribution, as well as on any number of other physical, chemical, electrical, optical, and magnetic properties. One multisensor approach using large, reliable, reproducible data bases mimics human operators to provide guideposts for integrating interrelated factors. For example, multiple sensors might provide information on forces, temperatures, and acoustic emission responses that can be processed in one or more ways (e.g., neural networks or parallel processors, least square regression fits to operating algorithms, or data handling by group methods), each intended to predict heuristically the value of a dependent variable from sets of independent variables.

Examples can be given of the role of sensors in the intelligent processing of AEMs. Piezoelectric transducers currently are used to pulse, receive, and process three independent, ultrasonic velocity components to provide in-process evaluation of the fiber-volume fraction, void content, degree of cure, and residual stresses of pultruded composites of fiber strands in a polymeric resin. When all bath control parameters and their interdependencies are fully established, quality improvement presumably will follow. Similarly, positional and mechanical sensors serve the multiple roles of process variable, control sensor, and recipient of encoded data and—with a laser beam—perform multilayer laser sintering to produce unique, controllable, and variable properties in turbine blade retrofitting applications. For less routine sensor applications (e.g., very high-temperature materials, unique combinations of materials with vastly different properties, and requirements for property controls at the atomic or molecular level), the technological barrier continues to be the lack of sufficiently predictable and controllable rule-based or knowledge-based systems to interface with existing and emerging sensors.

Technical Cost Modeling

Producers of AEMs must (1) understand the needs and desires of materials users and (2) provide desired products at the lowest possible cost. Although they seem simple in theory, these problems are ambiguous and difficult to solve in practice.

For example, the value of performance characteristics and the costs of achieving them are often not clear even to materials users. Moreover, explicit trade-offs, such as initial cost versus density versus stiffness versus service life, often are not well defined. Finally, numerous trade-offs and interactions among product design, raw material, and process will determine the cost of production once market needs become clear.

Methodological approaches to these problems, though still imperfect, have proved useful. One technique, multi-attribute utility analysis, has been used successfully to analyze the technical and economic trade-offs implicit in the materials selection process. Another, an engineering-based approach called *technical cost modeling*, uses computer simulation to determine the cost of producing components by alternative processing methods.

Here is one common formulation of the materials selection problem: Given a set of materials X, each possessing properties x, select the material X_i that, when used in product Y, gives the product the best set of characteristics $Z^*(x)$. The ease with which this problem can be stated belies the complexity of the computation required to solve it. The problem divides naturally into two parts. The first treats the problem of relating the materials properties, x, to the performance characteristics of the product, $Z(x)$. Though at present it may not be possible to characterize every performance characteristic of a product in terms of its constituent material, in general this part of the problem may be satisfactorily solved through a variety of engineering and economic analyses. The second part of the problem cannot be addressed so easily. Though engineering science is a remarkably complete tool for estimating the performance of a product as a function of its materials properties, it is ill-equipped to determine the best set of characteristics $Z^*(x)$. This latter problem can be grouped with a large class of decision analysis problems, termed *multiobjective problems*, that are characterized by the requirement that several objectives (or criteria) must be realized through a single course of action. In the case of the above problem, the selection of material X must lead to a best set of product performance characteristics that satisfy a potentially wide range of objectives.

The critical feature of this problem (and multiobjective problems in general) is that the values of the decision maker must be incorporated explicitly into the decision calculus. Only in extremely specialized circumstances can the best selection be made without an explicit treatment of the preferences of the decision maker. This point can best be demonstrated by a specific example of technical cost modeling (Figure 4-1). Suppose three alternative materials may be used to make an automobile body panel. A panel made of material A would cost \$10 to make and weigh 5 pounds; a panel made of material B would cost \$8 to make and weigh 8 pounds; a panel made of material C would cost \$9 and weigh 8 pounds. If the objective of materials selection is to reduce cost and weight, which material should be used?

Examination of the alternatives reveals that material C should never be used if it is always better to reduce cost and weight. Any panel made of material C could be made at less cost and at

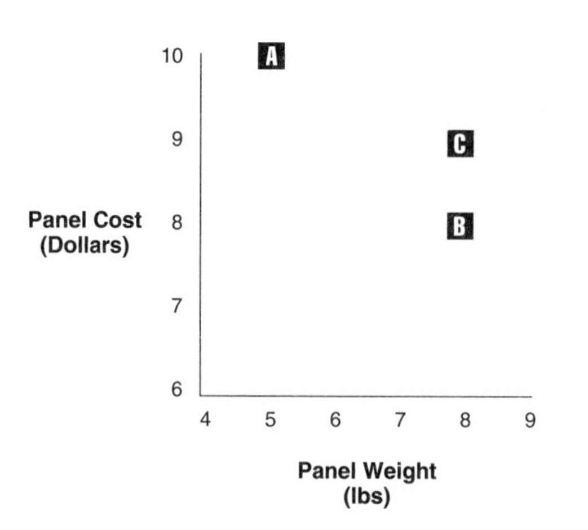

FIGURE 4-1 An example of technical cost modeling.

the same weight using material B. In the language of multiobjective problems, material C represents a "dominated solution" (i.e., other solutions are unqualifiedly better).

Of the remaining alternatives, material A offers low weight at high cost, and material B provides low cost at high weight. These alternatives are said to constitute the set of "nondominated solutions," solutions that, when compared to other members of the nondominated set, have characteristics that are at least as good and at least one characteristic that is better. Many other characteristics will have to be built into such a model to justify the use of expensive (boutique) materials, particularly if performance tradeoff is an option.

Technical cost modeling is an extension of engineering process modeling, with particular emphasis on capturing the cost implications of process variables and economic parameters. With cost estimates grounded in engineering knowledge, critical assumptions, such as processing rates and energy and materials consumption, interact in a consistent, logical, and accurate framework for analysis. Technical cost models can variously be used to:

- establish direct comparisons between process alternatives;
- assess the ultimate performance of a particular process; and
- identify limiting process steps and/or parameters, and determine the merits of specific process improvements.

Consider the results of a preliminary analysis, using technical cost modeling, of making a printed circuit board with an MMC. Currently, most high-powered avionics circuit boards are fabricated around a highly conductive core to facilitate heat removal, which must occur through the circuit board itself. Substituting for the current materials (usually copper-clad Invar (CIC) or copper-clad molybdenum (CMC)), an MMC such as graphite-reinforced aluminum could reduce the weight of the board without an appreciable loss in heat removal capability. To estimate the potential costs of such a board, a probable manufacturing process stream was developed and an associated technical cost model was used to explore the effects of manufacturing yield on the competitive position of the MMC-core board. The problem with manufacturing these boards is the low yield associated with laminating onto the metal core. The cost model was specifically designed to assess the effects of yield on the total cost of the board. As Figure 4-2 shows, the MMC board system must demonstrate yields on the order of 80 percent or higher to compete with the current CIC boards. Consequently, no such boards are being produced.

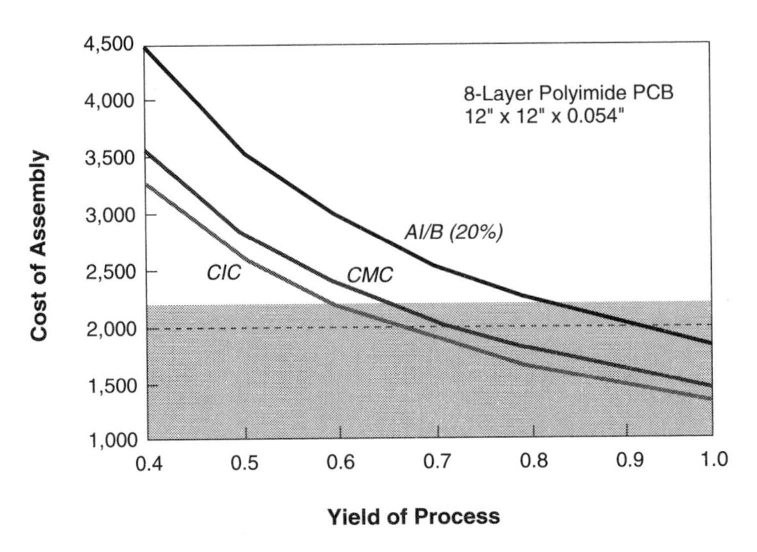

FIGURE 4-2 Cost of assembly versus yield of circuit boards with cores of copper-invar, copper-molybdenum, and aluminum-graphite.

Such economic models of both materials and associated processing are clearly critical for maximizing the usefulness of AEMs.

RESEARCH NEEDS

Research needs in AEMs fall into three overlapping areas: (1) materials-specific problems, (2) issues of integration, and (3) barriers to understanding resulting from inadequate education and training.

Materials

There is a pressing need for materials-specific research, particularly on the relationships among synthesis, processing, properties, and performance. Research in polymers, for example, should be directed at problems of fairly general or wide applicability, problems that, by their nature, will require the application of a variety of scientific and engineering disciplines. Examples include pultrusion with high viscosity thermoplastic matrices, effects of flow fields on discontinuous fiber orientation, and dispersion and adhesion of immiscible polymer melts.

The complex multiphase-flow conditions associated with the casting of MMCs, which limit exploitation of potentially low-cost foundry methods of manufacturing, also need to be studied. Molten MMCs, which may carry a 30- to 50-percent volume burden of fiber with a high aspect ratio, are very complex multiphase fluids whose dynamics and hydraulics differ significantly from those of molten monolithic alloys. Although these multiphase parameters are well developed and understood in the chemical processing industry, their adaptation to the foundry processing of MMCs is primitive at best. Research is needed to ascertain how existing hydraulic models can be applied and what new, more complex models are required.

The casting of slurries containing mixed solid and liquid metal and solid dispersed fibers, and the injection molding of a liquid matrix in molds containing long-fiber preforms, where penetration and partial solidification occur simultaneously, pose even more difficult problems. Both of these complex fluid flow processes, which are now treated empirically, will have to be better researched and modeled before they can be used successfully with MMCs.

Further, it is not obvious that the adaptation of powder metallurgy and foundry processes, the current methods of choice, to the manufacture of MMCs is most compatible with flexible manufacturing operations. Possible alternatives include liquid infiltration into long-fiber windings, tape methods analogous to those used for polymeric matrix materials, and vapor methods, including infiltration of fiber in vapor form into a porous matrix and infiltration of matrix vapor into a fiber preform. These and more radical processing approaches should first be modeled, then explored in the laboratory. Promising options could then become pilot studies.

Similar problems in the previous generation of AEMs (e.g., single crystal nickel-based superalloys) were solved by using interactive temperature and position sensors to extend empirical research to process models and control. Because property data are lacking, especially in multilayered heterophase structures for electrooptical applications, current problems tend to be much more complex.

In the future, it will be necessary to be able to design the microstructure to obtain a given set of properties and then design materials processing techniques to yield the desired parameters. Innovation in process development, using fundamental principles, will be needed to gain this objective. Methodologies for designing microstructures and processes to generate materials with desired properties must be developed as an academic discipline and taught to future engineers and scientists.

Integration

A focus on processing science is needed to create predictor maps to illustrate graphically the interdependence of materials, processing, manufacturing, performance, and cost. In addition, process models for representing thermodynamic and kinetic considerations need to be developed and integrated with realistic expert systems. Research is required to develop process modeling tools that relate processing conditions to materials characteristics and materials characteristics to materials properties and product quality. This research must yield a common modeling basis that can be exploited by simulations of disparate physical phenomena in order to arrive at better models of complex (fluid/heat) flow. The key to the success of this research is microstructural modeling and improved descriptions of microstructure. In a scientific climate that is marked by discontinuity between observation and theory, it is critical to extract maximum understanding from observed system behavior. Such is the case for materials processing, where data are extracted continuously from processing operations. It is clear that this expanding body of information can only be exploited through the use of expert systems.

Education and Training

Educational goals in materials processing and manufacturing science must be reformulated at both the undergraduate and graduate levels. The current lack of emphasis on fundamental processing principles and their relevance to the manufacture of products presents a problem over virtually the entire spectrum of AEMs, from composites to multilayer materials to electronic materials and beyond. A new focus on processing is needed, taught in one or more course sequences, to couple expanded knowledge bases in solid- and liquid-state diffusion, reaction kinetics, fluid flow, and other essential areas to a set of systematic manufacturing-oriented processing principles and practices that apply over a range of AEMs.

An alternative, or perhaps complementary, approach is the development of teaching factories. These joint ventures by academe, industry, and government would provide facilities for realistic materials manufacturing and processing experiments, which would undoubtedly lead to improved simulators, models, and controllers. The goal of this type of program would be to inculcate in students and practitioners the intellectual elements of an interdisciplinary manufacturing environment sufficient to educate them in new materials developments and manufacturing techniques. Individual teaching factories might be organized around particular state-of-the-art engineered materials technologies, reflecting the priority applications of the academic, industrial, and government partners.

The teaching factory concept might be extended to create a facility for rapid prototyping of advanced designs for or with AEMs or applications. The National Submicron Center at Cornell University and the Metal Oxide Semiconductor Implementation System (MOSIS) at the University of Southern California, for example, were initiated to support the fabrication of integrated circuit (IC) designs prepared at any university. The Submicron Center fabricates advanced electronic and optical structures by molecular beam epitaxy and e-beam submicron lithography. MOSIS, funded by the Defense Advanced Research Projects Agency, provides access to semiconductor process lines for fabricating IC designs in metal oxide semiconductor (MOS) technology for a given set of design rules that are updated frequently as the technology evolves. Designs are fabricated and shipped to their designers within weeks of receipt. The MOSIS program is said to have trained a new generation of IC designers. Yet another example is the Instrumented Factory Gear Center at the Illinois Institute of Technology, which is exploring new ways to produce, usually through machining, precision gears made of both conventional materials and AEMs. Materials properties are considered initially, and education and training are focused objectives of the program.

A host of similar opportunities can be envisioned for other AEMs and applications. For example, research programs currently exist that are aimed at producing flexible manufacturing cells and designing new materials and products. Creating a way to couple these design efforts and fabrication opportunities would greatly accelerate the use of new technologies and make more effective use of existing capital assets. A teaching factory might provide widespread access to rapid prototyping of designs for and with AEMs, and realistic, challenging products for flexible manufacturing cells. The resources of such a facility could be applied to a variety of materials processes such as: electrolysis of interface-dominated copper-nickel laminates (supermodulus materials), potentially useful in printed circuit boards or structural applications; new, dieless forming processes driven by computer-aided design (CAD) data typified by commercially available systems for plastic parts; compound semiconductor (e.g., gallium arsenide), molecular beam epitaxy (MBE), metal oxide chemical vapor deposition, or elemental semiconductor MBE systems (silicon and germanium), including, for the fabrication of semiconductor and structural products, an ultrahigh current implanter for oxygen and nitrogen; flexible turning or machining; and laser ablation requiring a facility capable of fabricating films from a broad class of materials systems, which provides a fertile field for new designs with dissimilar materials.

Although new facilities could be established as teaching factories, it would be advisable initially to test processes that could operate in existing facilities. Funding a facility to provide fabrication services would benefit the facility by providing data that could minimize process setup time and enable facility designers to assess factors such as design manufacturability. The installed MBE machines at university, industrial, and government laboratories, advanced IC processing lines such as those at SEMATECH, and the flexible turning/machining cells at university productivity centers are examples of existing teaching facilities, many of them very new as a result of university instrumentation programs and state efforts to create centers of manufacturing excellence.

The payoff from expanding such facilities and applying the teaching factory concept will be designers temperamentally geared to exploit AEMs early. The goal is to use untapped brainpower to fuel the discovery process in the hope of realizing new electronic and optical devices and new structural materials.

Research Guidelines

Research directed at improving understanding and promoting the use of advanced engineered materials should be guided by the following considerations:

- An early focus on integration of materials and processing requirements must be part of any manufacturing scheme.
- The effects of processing—both to produce the material and to produce the product—on subsequent properties and performance should be modeled and experimentally verified.
- Advanced structural and nonstructural materials, because of their unique micro- and macrostructures and ability to achieve highly desirable properties, should be central to the development of new manufacturing paradigms.
- Specific research needs include predicting and understanding interfacial and structural stability through processing and manufacturing steps, developing better process simulation methods that ac-

count for the microstructure of the material, applying appropriate AI and knowledge-based systems, particularly for processing, and obtaining real-time interactive control sensors.

- Manufacturing needs must be explicitly included in education and training materials for scientists and engineers. Toward this end the development of teaching factories to achieve some goals is highly recommended.

NOTE

1. ALADIN—The Aluminum Alloy Design Inventor is an AI system that aids in the design of new alloys. The system uses traditional, rule based, qualitative methods as well as quantitative calculations. It applies both Abstract Planning and Least Commitment Decision Making.

The problem of discovering a new alloy may be viewed as a search in the variable space defining an alloy. The initial state is a "root" alloy whose functional behavior is to be modified by altering the composition and processing. ALADIN uses symbolic reasoning to develop an abstract plan for the alloy design. This abstract plan contains decisions about the general microstructural features of the alloy to be designed as well as the alloying elements present and the processing methods uses. These quantitative decisions generate constraints on the quantitative decisions that are made later. A least commitment approach is taken to quantitative decisions, ranges of values are used as far as possible rather than single values. Determining whether these ranges and all the other constraints have a feasible region, i.e., if they can all be satisfied simultaneously, gives a criterion for the need for replanning.

Declarative structured representations are used for metallurgical data and concepts. The knowledge base contains information about alloys, products and applications, composition, physical properties, process methods, microstructure, and phase diagrams. The microstructure of alloys is described by an enumeration of the types of microstructural elements present along with their characteristics. This allows a representation of both qualitative and quantitative knowledge about the microstructure.

The methods developed in ALADIN forms a necessary basis and framework for future efforts focusing on specific aspects of materials design. For further detail see: "The Architecture of ALADIN: A Knowledge Based Approach to Alloy Design," by Hulthage, Farinacci, Fox and Rychener. IEEE Expert August 1990, pp. 56-73.

5

Product Realization Process

PRODUCT REALIZATION combines market requirements, technological capabilities, and resources to define new product designs and the requisite manufacturing and field support processes. The relevance and viability of specific elements of the product realization process (PRP) are determined by considerations related to the roles of (1) customers, including channels and suppliers; (2) technological feasibility, including information requirements; and (3) organization, including people, management, and the incentives and measures that affect productivity.

Customers

Corporate commitment to quality and responsiveness will be key differentiators in the competitive environment of the future. To accommodate rapid adjustment to customer needs, the PRP must view customers as integral to the organization. More responsive organizational structures and ready and rapid access to company information are essential for customers who want to be able to learn about the activities of and express their needs to companies in a timely manner. Customer requirements must be captured and made explicit—modeled, translated, and transformed in order to be useful in all phases of the product life cycle—and made available as needed throughout the enterprise.

Technology

The technological infrastructure must support the management of very short product life cycles, be able to satisfy customer expectations for improved quality, deal with cost-competition pressures, facilitate complex manufacturing processes, including the integration of frequently changing equipment technology and evolving manufacturing applications, meet demands for high equipment availability, and be able to handle enormous volumes of data. A technology architecture is needed that provides a modular, heterogeneous, distributed information processing environment that spans the entire system life cycle and includes specification of requirements, implementation, maintenance and enhancement, and evolution. Interfaces must be provided at all levels, from the factory floor (e.g., to processing equipment, robots, and materials systems) to corporate functions (e.g., to order entry, invoicing, and purchasing).

Corporate technology must not only support, but also must be the catalyst for interaction among all parties on the route from concept to product. It must foster the development and exchange of information between different corporate cultures, break down communication barriers, and shorten cycle times. The cost and complexity of advanced manufacturing technologies will usher in an era of cooperative ventures.

Organization

The need for prompt adjustment to rapidly changing demands in today's markets dictates an organizational framework for product realization that is flexible and adaptive. Development of such a framework involves reshaping the corporate self-image among employees at all levels. Employees must be ready to adapt quickly and know what adjustments to make. Two prerequisites for adaptability are open lines of communication and constant access to a wide array of information—in effect, an integrated data base. Knowledge sharing and adaptability are matters of policy and management. Line management must put communication and adaptation on a par with material productivity and timeliness, and performance must be continually monitored and refined to ensure that information is accurate and complete and that adjustments are effective and secure.

The commitment to communication must foster work groups and support outside business relationships. Work groups are needed that can draw on companywide information as the basis for decisions and immediately communicate their findings to interested groups. The needs and resources of vendors and affiliates must be reflected in the data base so that companies can immediately incorporate the results of new agreements in order to assess implications and make adjustments.

Conclusions

Shortening the product/process development life cycle will require both organizational and technological adjustments. Organizationally, it will involve the integration of product and process engineering into a single design effort that interacts with all aspects of the organization, including customers and suppliers. This organizational change will rely on, and must be made in conjunction with, the development of an integrated information systems infrastructure.

IMPORTANCE

The increasing frequency of new product introductions in some markets has reduced the life spans of some existing products. As the number of global competitors continues to grow and technological developments in materials, information systems, and manufacturing processes expand the scope of possibility, competition, previously cost-based, is becoming increasingly time-based—that is, driven by responsiveness to market requirements and time to product entry. Ultimately, as the coupling of the business relationship between producer and consumer becomes tighter, time-based competition will bring about an era of mass customization driven by the voice of the customer.

The emphasis in production has evolved from low-cost, labor-intensive manufacturing through capital-intensive, high-volume manufacturing to state-of-the-art, high variety, small-batch, flexible manufacturing. The latter reflects the need for increasing responsiveness to market changes and presupposes shorter product development intervals. The necessary changes in product realization—major organizational changes supported by major technological changes—are being driven by competition and fueled by technological advances.

Key external benefits of increased responsiveness include survival, first to market, more flexible market strategy (e.g., higher entry price), ability to use the latest technology, more accurate forecasts (resulting from shorter time horizons), and greater market share. Major internal benefits include control of product development through better regulated and communicated engineering changes, enhanced cooperation and more open information flow, and lower product realization costs (resulting from elimination of nonvalue-added activities).

VISION

As global competition intensifies, firms will increasingly seek differentiation from competitors through the PRP. Both organizational and technological concepts will be employed to accelerate the product life cycle. Such concepts as concurrent engineering, design for manufacturability, and business or *tiger teams* will become common practice. Tiger teams are cross-functional groups that are organized to address key business opportunities and are disbanded when results are achieved. In the future more tiger teams will be geographically distributed.

The PRP can be understood in terms of three planes: at the base, the physical plane (i.e., the tools, buildings, and physical processes and products); above that, the human plane; and at the top, the information plane (i.e., the concepts or representations of what is possible or desired). This is the model of product realization today—human beings working with limited information to create products through largely physical processes. In the vision for the future, the information plane assumes greater importance, human potential is amplified, and time dependence on the physical plane is reduced.

The resources of the information plane are constituted as product realization images. These images are representations in data of real world entities that behave as their physical counterparts do. Each image reflects the static and dynamic aspects of the reality it represents and can experience anything to which its physical counterpart in the real world can be subjected. Collectively, product realization images constitute an artificial reality that reacts to physical forces and breaks down or performs accordingly. But the greater power of these images lies in their ability to transcend the temporal limitations imposed in the physical plane; they can be used to accelerate time, to shorten the learning cycle, and to evaluate possibilities for trade-offs.

A complete set of product realization images would be determined by the set of products, processes, people, and business functions that interact in the course of the product development cycle. The product designer's image set would consist of product images that embody all key features and analyze and react appropriately to all physical forces that impinge upon them. The manufacturer's image set would consist of factory images—equipment and process images that contain the information needed to analyze the interactions among key features of product images. Marketing's image set would simulate end-user environments in which product images could be exercised, perhaps in a video-game-like manner. The complete image set would support any life-cycle interaction from conception to consumption. The simulation environment, by controlling the passage of time, would facilitate acceleration of the PRP.

Increasing opportunities to substitute information for traditional factory inputs, such as time, materials, and processing, will give rise to new organizational forms. Electronic virtual enterprises will use the evolving technology infrastructure to tap and coordinate the resources of a large number of geographically distributed organizations. Global information systems will enable networked firms to configure their product design, production, marketing, and distribution capabilities dynamically to produce low-volume products tailored to small, niche markets, as well as high-volume, mass-market products. These technological capabilities will have a profound impact on organization. Firms that take advantage of these capabilities will become industry leaders.

Computer-aided design (CAD) systems and emphasis on design for manufacturability will transform design practice. Feature-based design systems will enable designers to produce new products quickly. Such systems will evaluate manufacturability by using process, equipment, and materials models to validate product designs.

Advances in process control and materials composition will invite a reexamination of the way materials are transformed. Near net shape processes will form materials so precisely to shape that no cutting is necessary, thereby eliminating much waste. Near net shape processes such as hot isostatic pressing of titanium alloys for aircraft engine components and bulk growth of gallium arsenide single crystals of high quality and yield have demonstrated the viability of this concept. Computer simulations of process models will yield improved understanding of complex processes.

In the new product realization environment, physical capacities will become more like commodities. They will be represented by images in the information plane where organizational and technological transactions will occur much more rapidly. Companies will manage both internal and external relations in a constantly shifting pattern to maximize competitive advantage, being both competitors and cooperators on different contracts at the same time. New industrial relationships will introduce new organizational dynamics and new concepts of competition, ownership, nationality, resource availability, and profit allocation. Modeling in the information plane will greatly reduce the time constraints imposed by the physical world. Evolving product realization technology, adopted by organizations that have the capacity to change, will be the key determinant of success in the time-based competitive environment of the future.

PRESENT PRACTICE

Product realization, as it is practiced in companies with state-of-the-art organization and technology, can be viewed as a series of stages (Figure 5-1). In the text that follows, a definition is offered for each stage and the present practice is contrasted with a vision for the future. Overall, the vision for product realization is to reduce the time required to go from left to right in Figure 5-1, possibly even eliminating some of the stages.

Advanced Technology Development

Definition

Advanced technology development is the process of evolving new materials, processes, and tools for creating new, more competitive products. It is relevant research tightly linked to design, engineering, manufacturing, and marketing.

Present practice

Most advanced development organizations are at best loosely coupled to the rest of the enterprise. They tend to be autonomous organizations, separate from design, engineering, manufacturing, and marketing. Their inbred culture frequently presents barriers to communication with the rest of the enterprise and their mission statements are seldom tied to the ultimate success of a product in the market. Advanced development staff often do not understand where and how the new materials, technologies, and tools they are developing will be used.

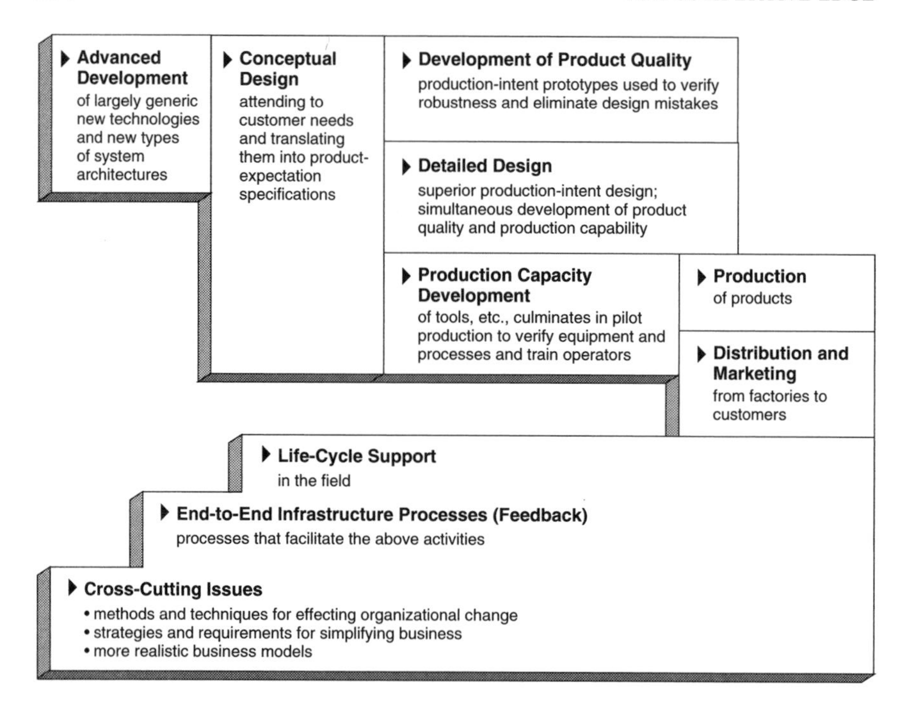

FIGURE 5-1 Product realization process.

Today's technology does not provide an integrated view of the information and intelligence that constitute a product image. Each organizational unit—from advanced development to marketing—tends to maintain its own data base and use its own language to describe a product. Much of the information generated in one stage is re-created in later stages with little or no integration. Negotiation among the various participants in the PRP is important. Taguchi's quality loss function approach offers one avenue for pursuing such negotiation.[1]

Vision

Improving the PRP depends to a great extent on having a full set of product realization images—representing advanced materials, tools, and technologies—that can serve as fast-cycle vehicles for technology assessment, feasibility testing, and technology transfer. As pressures of time-based competition force advanced development to reduce its cycle time, the ability to test concepts rapidly using a variety of scenarios will be a key competitive advantage.

Improvements in technology scanning, feasibility testing, and technology strategy development will enable many different product-technology morphologies to be tested, minimizing the need for costly and time-consuming development of physical prototypes and improving the quality of the end product.

Technology transfer will receive more emphasis as product life cycles and time to market contract. Manufacturing will be driven to understand and capitalize on every existing technology; fast-cycle competition will not allow time for retracing the steps of others or for reinventing a technology. This will give rise to the development and maintenance of a set of technology, process, and product images that will facilitate the transfer of collective knowledge.

Research in technology transfer would encompass the major phases, business processes, and roles of the technology transfer life cycle. It would cover the business and financial logic that governs the make-versus-buy decisions on which rest decisions on technology transfer versus in-house development. And it would explore ways of connecting advanced technology development with the realities of the market so that new products satisfy real needs.

Conceptual Design

Definition

Conceptual design involves capturing customer or market requirements and converting them to design specifications that are sufficiently detailed to permit concept selection.

Present Practice

The conceptual design process currently involves very little science and few methods or support tools to assure completeness and accuracy. Inadequately captured requirements lead to wasted resources, frequent re-engineering, missed market opportunities, and customer dissatisfaction. Those who contact customers, lacking marketing, technical, or listening skills, often inject their prejudices and biases into the process. Inadequate probing for detail may leave gaps or inaccuracies in the data obtained.

The concept selection process often lacks quantitative methods for assuring that the best business and technical decisions are made. Product team management has few formal methods and procedures for assuring clear and consistent communication of specific need and design intent. Similarly, few formal techniques or methods exist for quantitatively defining constraints, addressing feasibility management, accurately modeling product costs, engineering risks, or market opportunities, or selecting among alternative product concepts. Little research has been done on, and little scientific foundation exists for, the definition of comprehensive, testable, provably correct product specifications.

Vision

To gain the most from the conceptual design process a comprehensive framework will be in place, with tools that can integrate business, product, and market data to assure that optimal product selection decisions are made. Multiskilled teams will have formal methodologies for capturing complete and provably correct product specifications that can be tested against other models of the product.

Development of Product Quality

Definition

Product quality relates to the extent to which a product meets or exceeds customer expectations for functionality, robustness, and reliability.

Present Practice

A weak approach to the development of robust, reliable, manufacturable products is a major bottleneck in the U.S. product realization process. Prototypes that are built from nonrobust designs have much longer development times and, ultimately, result in inferior products. Research is greatly needed in both robust design and rapid prototyping. Current activity in the area of robust design consists almost exclusively of the implementation of methods pioneered by Box, Hunter, and Taguchi.

Vision

Robust values of critical design parameters are included in the first production-intent prototypes, which are available for testing soon after completion of the detailed robust design. Therefore, product robustness is developed concurrently with detailed product design. Designs are inclusive, spanning economic constraints to production feasibility. Research into robustness is multidisciplinary, involving both statisticians and engineers who understand the PRP. Optimization and learning, which today are largely disjointed, are integrated to achieve rapid, quality development. The optimization process not only integrates all available information, but also generates insights that will lead to continuous improvement.

Detailed Product Design Process

Definition

Detailed design is the process of translating customer requirements into a comprehensive description of a manufacturable product. It moves through the capture of design intent, to design verification, to the physical realization of a high quality, manufacturable artifact.

Present Practice

Design processes are often complex and not well understood. In the United States, life-cycle design (e.g., manufacturability, testability, installability) is still the exception rather than the rule. Furthermore, with the exception of drafting, design processes are still largely manual, with many handoffs and redo loops. Computer-aided engineering (CAE) has so far penetrated only about 10 percent of the engineering community. CAE tools are mostly stand-alone or poorly integrated, utilize closed architectures, do not generally reflect the unique processes of individual users, and tend to be aimed at specific engineering segments (e.g., IC design, circuit board and system design, and mechanical design). Though some university research is beginning to examine system architectures, most is still aimed at specific applications (e.g., simulation and design compilers/synthesizers).

Vision

In the envisioned product realization context, designers and other members of the product realization team will work concurrently, in a paperless environment, with minimum queue time and rework. Tools will be flexible, enabling the design process to dictate the use of the tools rather than vice versa, and allowing designers to move easily from one design environment to another (e.g., from electrical to mechanical, from IC to system level). Design systems will have greater intelligence, allowing designers to perform trade-off analyses in order to optimize designs for life-cycle needs. These tools will also assume many mundane engineering activities, such as schematic capture and prototyping, thereby allowing engineers to move to higher levels of abstraction and architectural design where leverage is greater and where 80 to 90 percent of life-cycle costs are determined.

Production Capacity Development

Definition

Production capacity development encompasses process planning; design, quality, tool, and factory industrial engineering; facilities planning; training; and production staffing.

Present Practice

In most industries, manufacturing, marketing, and product design are separate management areas that often have conflicting goals and communicate very little with one another. As it stands today, the many systems that support these individual areas cannot be integrated to provide a seamless view of the capabilities of the organization.

Vision

A need exists for close coordination among design, manufacturing, and marketing departments to produce good models of production capacity that can be integrated with models of material flow and process capability.

All the information needed for production capacity analysis, though highly detailed and existing in different systems, will be online and available. Organizations will make information available wherever it is needed throughout the product life cycle, and they will have the data integration and communications capabilities that will enable them to do so.

Production

Definition

Production is the means by which raw materials are transformed into products of appropriate quality at minimum time and cost. Production assumes a detailed product definition, including a description of function, geometry, materials, and tolerances, a process plan, and production capability.

Present Practice

Production delays are often caused by engineering design that optimizes functionality and ignores the impact of design on later stages of the product life cycle, such as fabrication, assembly, testing, distribution, field service, and reclamation. The extent to which a design is feasible from a cost and quality perspective is limited by the available set of processes; poor choice of tooling may lead to inefficient use of production facilities (as recent experience with flexible manufacturing systems has demonstrated), and designs that specify too many parts can increase both the cost and complexity of assembly.

Control of production processes has become more complex with the introduction of programmable automation and the greater flexibility of machining centers and robotics. Yet existing control systems are barely able to manage the factory of the past, let alone the factory of today or of the future. Resource competition also complicates shop floor control. Bottlenecks caused by competition for resources are a major impediment to productivity, and batch sizes continue to be reduced to meet rapidly changing demand, exacerbating the difficulty of allocating resources. Finally, the ability to control production is limited by the accuracy and timeliness of information from the factory floor. Current information acquisition systems, designed to support accounting-related tasks, do not meet the information needs of a dynamic control system.

Production knowledge is limited by the artificial separation of engineering and manufacturing. Traditional principles that exhort managers to break down organizations into pieces that are more easily managed have led to the separation (logically and often geographically) of engineering and manufacturing groups. This separation has impeded communication and coordination to the point that the two groups no longer speak the same language.

Vision

The ability to adapt rapidly to new materials and processes or new knowledge in engineering and science ultimately will reduce production costs while simultaneously increasing product quality. Consequently, next-generation control systems (as explained in Chapter 2, Intelligent Manufacturing Control) will be highly flexible, able to analyze the production situation, and make the best control decision in view of current goals, opportunities, and constraints. These systems will also monitor their own performance, at both the unit process and shop levels, identify poor performance, and diagnose and eliminate the cause. As production systems become more complex diagnoses will be based on deep models of the process, necessitating reliance on model-based reasoning.

Communication will extend beyond the factory floor. Production managers will communicate, coordinate, and negotiate with earlier and later stages in the product life cycle, negotiate changes in product definition to optimize production cost and quality while design is still under way, and contract with suppliers to guarantee availability of the necessary materials and parts. Production managers also will have the latitude to identify and communicate to other stages of the life cycle significant production events, such as quality problems caused by design decisions, customer feedback, and consequences of vendor/supplier interactions. Such integration will be necessary whether production is tightly coupled to colocated product life-cycle stages, such as engineering, within the same firm or is done in a separate facility that works with other firms.

Production will have more powerful management systems that acquire, filter, and communicate significant information to those who need it. This capability will be applied within a dynamic, real-time environment, necessitating the existence of a model of the factory that is precise, accurate, and realistic. The speed with which new materials, processes, and products are introduced will leave little time to analyze them thoroughly in order to optimize production quality. Optimization of production quality will, instead, be achieved over time by new tools for continuous improvement of process models and control strategies.

Distribution and Marketing

Definition

The concept of distribution and marketing includes selling products quickly, responding continuously to customers' needs and desires, making the public aware of the variety of products available, and distributing products to customers.

Present Practice

Distribution and marketing are separated both organizationally and technologically from the rest of the product realization cycle. As separate organizations, they lack both access to information and the influence necessary to affect product realization times.

Vision

Just-in-time (JIT), a method of manufacturing by which parts and assemblies are made or delivered as needed, thereby greatly reducing inventory, was initiated as a process for the factory and its suppliers. It will be exploited in the consumer distribution system. Marketing will be revolutionized by the development of a preference-indicating computer that will enable customers to view product options on a color screen that displays revolving and cutaway views, specifications, and other information, and that is updated weekly, or perhaps daily, to include the latest product improvements.

Life-Cycle Support

Life-cycle support, which must accommodate output ranging from lot sizes of one to mass production in a global marketplace, includes open architecture (both business or organizational, and systems), continuous quality improvement, field serviceability and maintainability, continuous cost/performance improvement, and end-of-life planning. Open architecture affects all aspects of life-cycle support, from product conception through the product sunset decision. Organizational changes are required to support an open architecture framework, in which dynamic product configuration will be used to respond to customer change requests within hours rather than days, weeks, or months.

Promotion and use of international standards will reduce both the development and long-term sustaining costs of product realization; modularity will reduce sustaining efforts. Development of an infrastructure or framework composed of reusable components and off-the-shelf, plug-compatible modules will provide faster product-to-market transition and reduce ownership sustaining costs.

Embedded diagnostics, standards, and development and implementation of fault-tolerant and error-correcting designs will increase product reliability, thereby reducing product maintenance costs. Field service and maintenance costs should be monitored as part of a continuous cost/performance improvement methodology. These data should be part of models that feed directly into advanced development. (More on these practices can be found in Chapter 3, Equipment Reliability and Maintenance.)

The PRP also must include a plan and process for end-of-life planning. The manufacturing business process must account for manufacturing material waste and develop cost models that provide insight into the total cost of ownership of the product. Cost-of-ownership models should include discontinued availability, archiving, recycling, and obsolescence planning for product and process.

End-to-End Infrastructure Processes

The rate of product development and improvement is closely related to the rate and cost at which feedback can be obtained and implemented. Qualitative and quantitative analysis, simulation, prototyping, and testing are key feedback mechanisms used throughout the product life cycle.

A unified design language capable of representing design intent as well as designed objects at various levels of abstraction (from functional description to 3-D geometry) would greatly improve the productivity of people involved in product realization. Because many products are designed and developed by large project teams, simultaneous information sharing among team members is important. As only a comparatively small part of the knowledge content of the PRP will be describable in analytical terms, the availability of tools for the efficient capture, representation, and verification of knowledge is essential. Simulation is as important a tool for verifying design ideas and concepts as it is for furthering intelligent manufacturing control, improving equipment reliability and maintenance, and supporting manufacturing of and with advanced engineered materials. (See Chapters 2-4.)

The practice of making product development decisions sequentially frequently leads to many iteration cycles. Design infrastructures that support multiple and simultaneous views of information allow designers to check for consistency and design integrity as well as eliminate some of these iteration cycles. Experience gained during product development that might benefit the development of similar products should be subject to capture in a continuously evolving knowledge base. Such knowledge bases should greatly reduce the time required for trial and error cycles and ultimately may be able to adapt dynamically to trends and changes in manufacturing technology and the market.

Managers, as well as product developers, stand to benefit from the product realization infrastructure. Management will require additional decision support tools, such as risk and value analysis, as well as the ability to monitor the stages (check points) of product development. Management may come to view a system that can perform these functions as a kind of "war room" terminal.

As business requirements change, necessitating new organizational structures, the PRP must be readily reconfigurable. This adaptability will rely heavily on a computer-integrated enterprising (CIE) architecture that supports maximum individual input and responsiveness. Such an architecture only will be effective in the wake of organizational realignment that supports the free flow of information and rapid decision making.

Cross-Cutting Issues

The PRP today is for the most part serial, with each stage of design performed independently and sequentially. The organization is modeled after the design process. The lack of information in a readily available format stifles change. Accessibility of information can be enhanced by various methods and techniques (Figure 5-2).

RESEARCH NEEDS

The ability to represent on the information plane the set of product development activities that exist on the physical plane is the key to more efficient product realization. Research in this area should be directed at developing the capabilities and infrastructure needed to achieve the vision proposed in this report—in which a set of artificial realities (intelligent images of product, factory capabilities, customer specifications, and organizational structure) with the ability to interact functionally (e.g., a product image capable of communicating with a factory image about its manufacturability and considering trade-offs between product performance and manufacturing, inspection, and maintenance) is allowed to play out in a suitably structured environment. Attainment of this vision relies on defining, identifying specific instances of, and developing intelligent images; identifying and establishing the requisite connections among these images; and devising an organizational structure in which these concepts can be made operational.

ADVANCED DEVELOPMENT
Completely integrated with product design, manufacturing, and marketing. Supported by simulation and technology transfer.

CONCEPTUAL DESIGN
Formal methodology for gleaning customer specifications and integrating them with existing product data.

PRODUCT QUALITY
Product robustness is an integral, quantifiable part of the design process.

DETAILED DESIGN
Intelligent design systems and cooperative design teams will permit a focus on solving theoretical design issues, and resulting cost savings.

PRODUCTION CAPACITY DEVELOPMENT
Coordination among product design, manufacturing, and marketing managers will facilitate complete production capacity analysis.

PRODUCTION
Will integrate data from all stages of the product life cycle.

DISTRIBUTION AND MARKETING
JIT used in distribution, and a preference-indicating computer will function as a customer catalog (new marketing concept providing greater detail and periodic updates).

LIFE-CYCLE SUPPORT
Revamp at all stages using open architecture, international standards, module framework, diagnostic approach to product maintenance, end-of-life planning.

END-TO-END INFRASTRUCTURE
Cyclical rather than sequential view of PRP.

CROSS-CUTTING ISSUES
More integration of design process, with an emphasis on information sharing.

FIGURE 5-2 The product realization process of the future.

Intelligent Images

Product realization as envisioned above is CIE in the most fundamental sense: total integration of CAE, CAD, and computer-aided manufacturing (CAM) systems with systems for developing and communicating customer product specifications, all operating in a simplified, streamlined process. Considerable research in data capture and management will be needed to develop the necessary interactive, intelligent product realization images. These images would be ensembles of data, algorithms, and geometrical, mathematical, and empirical multimedia models that capture the character and intent of the various agents and activities that comprise the product development life cycle.

Specifically, research is needed to define the nature of an intelligent image and identify the set of images that will be needed to complete the product development cycle in the information plane. A product image is clearly indicated. A factory image that expresses the manufacturing capabilities and constraints of the enterprise, and a customer image that gives rise to and interacts with the product image, also are suggested.

Each intelligent image must provide multiple views to accommodate the various perspectives of participants in the product development cycle. For example, design components organized by engineers according to functionality will be viewed by manufacturing managers as production constraints. Similarly, process planners will focus on different product attributes than will those concerned with functional simulation. An intelligent image must be representationally rich enough to support the construction and maintenance of multiple views without incurring the overhead of multiple data bases.

Research also should be directed at the development of shareable engineering and manufacturing models whose contents are precise and clear. Examples include research on axioms of deduction based on the ontologies of engineering and manufacturing knowledge, and interfaces for the creation, perusal, and alteration of semantic models.

New methods and technologies are needed for capturing, organizing, and managing requirements data for product realization. Methods must be developed for documenting, communicating, and integrating throughout the product development cycle design intent derived from original customer requirements. This effort presupposes a better understanding of the overall process of design. Research should be aimed at determining how the various subfunctions of design might be represented informationally, how the required flexibility might be provided, and how the complexity implicit in design and manufacturing information might be accommodated. It also should be aimed at determining whether there is a fundamental trade-off between rapid access and inherent flexibility that evolves to a problem of managing bandwidth.

Ways must be developed to capture, easily and precisely, strategic and tactical business and manufacturing requirements and to model, store, display, and transform these requirements into readily available automation and manufacturing solutions. Developing capabilities in this area will rely on research into (1) methods for documenting, communicating, and verifying design intent and customer use intent at each stage of product development; (2) models for capturing and transforming customer preferences into multiple views for design and manufacturing; (3) systems for validating requirements for internal consistency, integrity, and completeness; and (4) display technologies capable of communicating the full richness of intelligent images and of integrating data with functional, user, and control models.

Linkages and Relationships

Interaction is as essential as intelligence in the product realization images suggested as the basis of future product realization. An intelligent image would need to understand not only itself, but also relationships and transformations of information in different stages of the product development cycle. A product image, for example, would need to understand not only its specifications, composition, characteristics, and manufacturing requirements and constraints, but also its transformation into the manufacturing capabilities expressed by the factory image and the feedback from that image on its manufacturability. Both of these images would need to interact with the customer image to determine, for example, what changes to the product image within the constraints of customer specifications might enhance manufacturability or reduce production costs. Intelligent images would be required to anticipate the information needed at all stages of the product life cycle and determine its optimal form and content.

The supporting technologies needed to provide this capability include high performance application tools such as prototyping tools and simulators, design compilers and synthesizers, rapid imaging techniques, and trade-off and selection analysis tools based on AI and expert systems. They include links to production such as rapid part formation through, for example, holographic polymerization, contourable dies, and particle deposition. These technologies will demand environments that provide complete and accurate sources of information for analyzing and optimizing designs, such as semantic and object-oriented data bases, simulation models, and component reliability data. Research should aim at identifying techniques for developing a comprehensive factory/product data base capable of integrating CAD-developed point solutions for tooling requirements, cost accounting, manufacturing requirements planning, capacity planning, and factory layout modeling into an integrated view of production capacity. Such an environment would serve to automate the review and approval process among distributed peer work groups and accumulate valuable experience from product development in a continuously evolving knowledge base. Here, research is needed into the development of efficient tools for capturing, representing, and verifying knowledge.

Finally, systems are needed that explain the hidden costs of manufacturing and existing long product-development intervals and relate them to the forces responsible.

Organizational Framework

Alternative management structures must be developed that are capable of preserving product integrity and quality while coping with rapid market and technological changes and their impact on corporate culture. The organizational structure required to manage the linkages and relationships by which intelligent images accelerate the product life cycle—from design and development, through manufacturing, to marketing—is a new beast that must live in a dynamic environment. Its creation involves reshaping the corporate image among employees at all levels and flattening the organization to allow more localized decision making. Rapid access to information is of little value if people cannot act on the information. Attuning employees to the corporate network is a challenge relating to role definition and motivation that classical hierarchical management does little to address.

Altered conceptions of manufacturing and work organization almost always accompany major shifts in the technology of production and appear to be essential to the successful use of new technologies. It is precisely in change-related conception and execution that many U.S. firms badly lag behind their Japanese and European counterparts. Comparison of European and Japanese response to changing technology suggests no single model of successful execution. The firms conceive and organize production, using essentially identical technologies, in substantially different ways and achieve varying levels of success that do not correlate with the particular approach taken. Despite these differences, European and Japanese experience shares an important feature—truly remarkable leaps in manufacturing performance and productivity appear to occur only when new production technologies are applied in changed industrial settings.

In contrast many more U.S. than European and Japanese firms allow technology to drive organization and conceptual change. Whether this strategy can succeed is a very real question. Systematic study is needed to determine how firms in different countries and industries conceptualize and organize manufacturing, and to identify variations in the practices they use to implement manufacturing technologies like those described in this report.

U.S. industry is approaching consideration of all product life-cycle requirements early in the design cycle by formation of multidisciplinary teams. Because it involves handling people, the team concept has been viewed as a management issue when, in fact, it is much more. Some of the most interesting research problems lie in group design processes. Issues range from user interfaces that enhance consensus building to life-cycle models that interface with each level of abstraction in the design process to the information architectures needed to bring all this information together in a unified, responsive whole.

To ensure collegiality among knowledge workers who must collectively attack a problem, the twenty-first-century approach may exhibit mixtures of the sciences with organizational development to meet the needs of changing life styles. Such is the concept of the village industry, which supports interaction and camaraderie among clusters of experts while maintaining proximity to the family. The "village" will likely be closer to experts' homes than to company facilities, and performance will be measured on a deliverables basis, rewarding teams instead of individuals. Significant research issues—such as candidate identification relative to project type, and type of participant—must be resolved, as must other issues, such as technology access and security. Additionally, measurements for team rewards, present in few companies today, must be developed.

Summary

As more of the product life cycle moves onto the information plane, organizational structures and work norms must become more fluid. Any computer workstation might become the workplace, and the organization itself will be a flatter structure that relies on electronically networked peer-to-peer business teams that are readily reconfigurable to respond to rapidly changing business requirements. The PRP, in maturity, will catalyze the development of a new kind of manufacturing—the manufacturing of complete product images that are as marketable as the physical products they represent. In doing so, it will usher in the true information economy.

The research that will enable U.S. manufacturing to reach this plane lies in six general areas:

- Definition of, and supporting technology for, intelligent images;
- Data base structures that combine engineering, manufacturing, cost accounting, capacity planning, and factory layout modeling into an integrated view of production capability;
- Technologies for capturing, organizing, managing, and displaying data on strategic and tactical business and manufacturing requirements;
- Group design processes that support and enhance the functioning of multidisciplinary work teams;
- Management that eliminates the dichotomy between flexibility and productivity; and
- Control architectures that facilitate movement up the intelligence ladder beyond feedback, memory, and learning to goal changing.

NOTE

1. Taguchi, G. 1988. Introduction to Taguchi methods. Engineering 228:1-2. (The methods of quality improvement developed by Genichi Taguchi have already found widespread acceptance in Japan and the United States. Based on a different way of thinking about quality, these methods use statistical analysis to ensure high product quality. Taguchi's thinking on quality is based on two fundamental concepts: that any loss in quality is defined as a deviation from a target, not a failure to conform to an arbitrary specification; and that high quality can only be achieved economically by being designed in from the start, not by inspection and screening. Taguchi's definition of quality is customer-oriented. Quality is the characteristic that avoids loss to society after the product is shipped. A loss of quality can therefore be measured in pounds, dollars, or yen. His philosophy is that adding features is not a way of improving the quality of a given product, only of varying its price and its target market segment.)

6

Manufacturing Skills Improvement

ADVANCED MANUFACTURING technology will transform the image of manufacturing employment from a sweaty job of last resort to an intellectually demanding occupation.

In making knowledge an implicit part of manufacturing practice, for workers as well as management, advanced manufacturing technology is creating a need for a more educated work force. The shift in educational attainment by manufacturing employees between 1973 and 1983 (see Figure 1-4, p. 9) will become more pronounced in the coming decade. This shift is occurring at the same time that the supply of potential young workers is beginning to decline precipitously (Table 6-1, Figure 6-1). For example, in Germany in the year 2000, the pool of young workers is expected to be 60% of what it was in 1984. This decline is compounded by the fact that many economically disadvantaged individuals cannot meet even minimal skill requirements for the new manufacturing jobs.

The development of manufacturing skills does not occur in the abstract. It is related to a set of goals, specifically to the creation and maintenance of a well-trained, flexible, and motivated manufacturing workforce, comprising prospective workers as well as current workers at all conventional levels, including technical professionals and managers, mid-level technicians, and shop floor personnel.

Education of prospective manufacturing workers typically occurs in elementary and secondary schools (grades K-12), in community and technical colleges and trade schools, and in professional colleges

TABLE 6-1 Size and Ethnic Distribution of 22-Year-old Population, 1980-2000.

		Percent Distribution				
Year	Total (000s)	White	Black	Hispanic	Asian	Native American
1980	4315	77	13	7	2	1
1985	4213	76	13	8	2	1
1990	3601	73	14	10	2	1
1995	3346	71	15	11	3	1
2000	3350	70	14	12	3	1

SOURCE: E. L. Collins. 1988. Meeting the scientific and technological staffing requirements of the American economy. *Science and Public Policy* (15:5): 335-342.

in engineering and business. The current work force usually is trained—typically in basic skills, communication skill, and skills related to teamwork and group dynamics—through continuing education and training and retraining programs. Retraining tends to be job-, industry-, or company-specific and to be structured by levels (e.g., upper, middle, and lower).

This chapter examines education and training for all manufac-

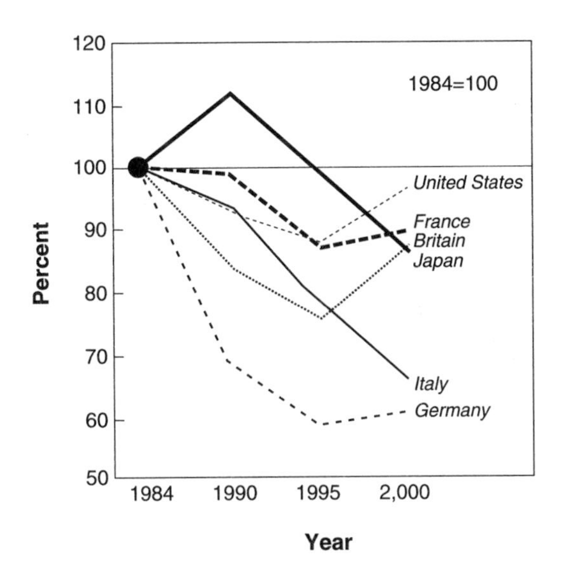

FIGURE 6-1 Index of supply of potential young workers (15-19 year olds), 1984-2000. Source: Institute of Manpower Studies, International Labour Office.

turing workers at all levels. It focuses on the skills needs generated by the technology described in this report. Delivery mechanisms—such as workplace training and college-industry coalitions—are discussed in relation to both current and prospective workers. Finally, the chapter recognizes, but does not address, issues related to basic literacy, general education, and the need for broad changes in attitudes toward the importance of manufacturing.

IMPORTANCE

A 1986 study found that more than 40 percent of the work force operating advanced manufacturing systems in Japan were graduate engineers and that the remainder were technically well qualified. That the U.S. work force is not as well qualified is only part of the problem facing U.S. manufacturing firms. Also in serious doubt is the ability of the country's prospective work force to meet the skill requirements imposed by the advanced manufacturing technology on which our international competitiveness depends.

As mentioned in Chapter 1, *The Economist* has reported that 6 out of 10 of the nation's 20-year-olds cannot add up a lunch bill.[1] *The Wall Street Journal* has reported that 58 percent of Fortune 500 companies complained in a survey of having trouble finding employees with basic skills.[2] Southwestern Bell, according to the *Journal* article, in 1989 processed more than 15,000 applications to find

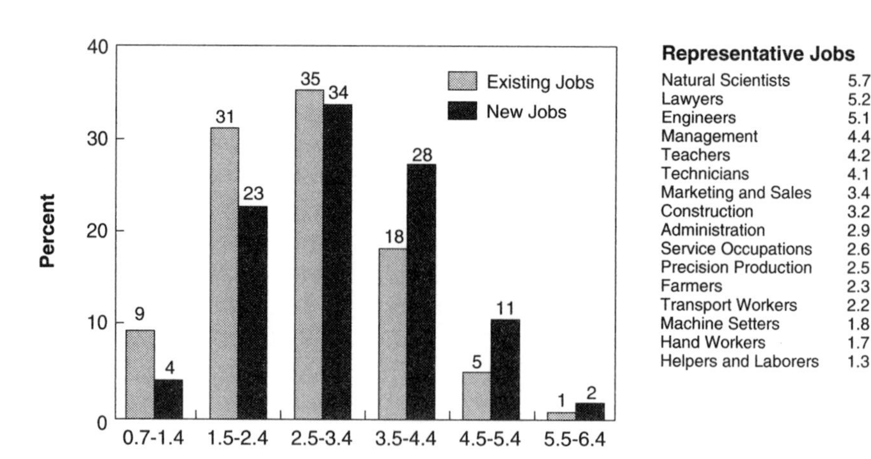

FIGURE 6-2 Percentage of low-skill jobs is declining. Source: W. B. Johnston and A.H. Packer. 1987. Workforce 2000: Work and workers for the 21st century. Hudson Institute. xxi-xxii.

TABLE 6-2 Changing Educational Requirements, 1984-2000.

Level of Education Required	1984 Jobs	2000 Jobs
8 years or less	6%	4%
1-3 years of high school	12%	10%
4 years of high school	40%	34%
1-3 years of college	20%	22%
4 years of college or more	22%	30%
Total	100%	100%
Median years of school	12.8	13.5

SOURCE: W. B. Johnston and A.H. Packer. 1987. Workforce 2000: Work and workers for the 21st century. Hudson Institute. 97-98.

3,700 people to take its 34-minute basic skills test (e.g., mathematics questions that call only for the operation to be identified, not for the computation to be performed). Only 800 passed, and further screening, in the form of interviews, physicals, and drug tests, resulted in only 580 new hires, at an estimated cost to the company of $1,000 per job.

The educational crisis reflected in these circumstances is of national proportions. A Hudson Institute study,[3] furthermore, predicts that more than half of all jobs created between 1984 and 2000 will require some education beyond high school and almost a third will be filled by college graduates (Figure 6-2, Tables 6-2 and 6-3).

Department of Labor projections for manufacturing employment to 2000, though level, show a change in occupational mix, with greater proportions of engineers, technicians (who may be upgraded operators), and managers (see Appendix A).

TABLE 6-3 Fast-growing Jobs Require More Language, Mathematics, and Reasoning Skills.

Rating (scale of 1-6 with 6 the highest)	Current jobs	Fast[a] Growing	Slow[b] Growing	Declining
Language	3.1	3.8	2.7	1.9
Mathematics	2.6	3.1	2.3	1.6
Reading	3.5	4.2	3.2	2.6

[a]e.g., professional, technical, managerial, sales, and service jobs.
[b]e.g., machine tenders, assemblers, miners, and farmers.

SOURCE: W. B. Johnston and A. H. Packer. 1987. Workforce 2000: Work and workers for the 21st century. Hudson Institute. 98-99.

Widening of the gap between the growing demand for engineers and managers and the supply of appropriately educated and trained individuals will occur for four main reasons. One, as mentioned earlier, the number of new labor market entrants is declining, reflecting a decline in the population of 16- to 24-year-olds. Two, the growing numbers of women and minority persons who will be needed in the manufacturing work force to offset this decline do not have the educational base to acquire the necessary skills. Three, the low status of manufacturing as a career continues to dissuade many qualified people from pursuing careers in this field. And finally, increasing deployment of robots and other automated processes will further widen the gap by rendering many lower level manufacturing occupations as irrelevant as science and engineering are indispensable. The supply of manufacturing teachers, at all levels, will be affected by these same issues and reflect the same patterns.

A Department of Defense (DOD)[4] report summarized existing deficiencies in manufacturing skills and skills acquisition. The report faults the teaching of management theory and practice in the United States, summarized by statements such as, "good management is management by financial control"; "good managers can manage anything"; "individual achievement is important, not teamwork"; and "manufacturing is an unimportant function." This approach, it says, is to blame for the inability of U.S. managers to achieve manufacturing results equivalent to those achieved by their Japanese counterparts. The report also reprimands engineering schools in universities for training engineers for careers in product research and development at the expense of an adequate focus on manufacturing. Few faculty members, it says, have industrial experience or expertise, and emphasis on specialization produces engineers who are ill-equipped to understand total manufacturing systems.

The report notes in addition a severe shortage of adequately trained scientific and engineering students. In this vein, Dean Meyer Feldberg of Columbia University Business School observed in an interview with *The New York Times* that fewer than 4 percent of all college students will graduate in engineering, compared to 24 percent who will graduate with degrees in business.[5] Dean Feldberg noted that Japan, with half the population of the United States, graduated twice as many electrical engineers as this country in 1989. Of the graduate students in science and engineering in the United States, almost half, according to the DOD report, are foreign. The report cites inadequate industry programs of continuing professional education and training for engineers and production workers in the existing work force. A number of leading firms that have established training programs to upgrade the skills of production professionals, technicians, and operators report significant gains in production and increased ability to attract first-rate people to production jobs, but the bulk of U.S. industrial firms have neither the money needed to develop effective programs nor access to instructors competent to teach a broad range of modern manufacturing skills.

Apprenticeship, despite criticism and unfavorable appraisals by academics and public policy analysts, "has stubbornly persisted and actually thrived in certain occupations, industries, localities, and countries," according to a report from the National Center for Research in Vocational Education.[5] The report reveals that a rich diversity of apprenticeship practices exists in the United States unbeknownst to many for want of effective means of fostering awareness, dissemination, and replication. Comprehensive final examinations for apprentices, though common in other countries, are rare in U.S. programs, according to the report, and state and federal support for the largely privately sponsored and financed programs is incomplete and uncoordinated. Although some public funding of apprenticeship programs does exist, primarily for curriculum development and related classroom training, industry financing through training trust funds, sometimes jointly administered with unions that have a strong interest in training, is more common.

In addition, a number of institutions and organizations in the public and private sectors, many with overlapping constituencies and missions, provide education and training for manufacturing that covers pretechnical, entry-level, technical, professional, skill upgrading, and continuing education. These agencies employ a variety of means and methods, ranging from the informal (e.g., on-the-job training) to the exotic (e.g., satellite-transmitted professional courses). Training for manufacturing is not standardized across the industry. Students and trainees may encounter full-time faculty or part-time instructors drawn from industry. The experience of these teachers may range from extensive to nonexistent; faculty may lack technical knowledge, and industry instructors may lack teaching skills. Many will be unaware of the availability of models, curricula, guides, and other teaching resources because no central clearinghouse for such materials exists.[6]

If U.S. economic well-being, quality of life, and national security are to be maintained in the face of a contracting working-age population, those who are capable of working will be required to work smarter and harder. They must be equipped to do so. If they are not, the United States stands to lose both the ability to create technology and the ability to absorb technologies created abroad.

BARRIERS AND CHALLENGES

The improvement of manufacturing skills in the United States faces two general barriers. One is the lack of resolve regarding manufacturing and its importance to the national well-being. A shared national value, such as the importance of education and national defense, arose for manufacturing only briefly, during World War II; its renewal could serve to enhance understanding of manufacturing and so improve its image.

Manufacturing's image problem derives from an outdated, circa World War II, perception of manufacturing jobs as dirty, hard, low paying, and confining, as well as from a more contemporary lack of understanding, among many managers as well as government officials and the general public, of what manufacturing entails and of the value of manufacturing careers. One result of this image problem, according to the 1988 DOD report, is that manufacturing does not compete effectively for high-quality personnel. A business school dean observed recently that the same companies that send vice-presidents of marketing and finance to recruit for those functions often send a personnel specialist to recruit for manufacturing jobs. "The salary structures they offer," the dean remarked, "might as well have a big sign attached that says: 'Don't apply for this job category.'" Until this image is changed and public understanding enhanced, people not previously motivated to choose manufacturing careers can hardly be expected to do so.

The other general barrier to improving U.S. manufacturing skills is the lack of a coherent national policy and standards for human resource development. This shortcoming has implications for all teachers in elementary, secondary, trade, and postsecondary schools, as well as for industries, vendors, and labor unions. These absences, of a national will toward manufacturing and a national human resources policy, confound efforts to surmount the more localized barriers identified below.

Part of the challenge of improving manufacturing skills is to identify them. The increasing sophistication and rapid change in manufacturing processes call for higher level skills that can be continually augmented, adapted, and modified. Advanced manufacturing technologies have created a need for work teams that are highly skilled not only in job-specific and general technical abilities, but also in interpersonal skills and organizational management. Differences in specific skill requirements between small and medium-sized companies and large firms tend not to be reflected in existing curricula, whose development and adoption is most often driven by the views of large firms. Customized training, long a staple of two-year colleges, is aimed almost exclusively at the needs of large firms, which are often those that can best afford to provide their own training.[7]

In the present U.S. manufacturing context, skills are generally expected to come from elsewhere. Industry pays taxes that in part support an educational system that it expects to meet its needs, and society is largely content to leave skills acquisition to individual initiative. Many potential providers of training in manufacturing are inhibited by lack of resources or motivation. Small and medium-size companies typically lack the time, money, and competence to mount effective training programs, while schools funded on a full-time equivalent basis often cannot justify programs designed to train small, dispersed enclaves of students. Given the mobility of the work force, many companies are reluctant to invest in the training of workers who might take their new skills elsewhere. Finally, people do not properly appreciate the benefits of training. Part of this problem derives from the seeming inability of U.S. business management to adopt policies that would allow companies to pay for skills (as opposed to jobs). Evaluations of apprenticeship programs, for example, seldom take into account the benefits that might accrue to an employer if an apprentice stays with the firm after completing the apprenticeship. But very real difficulties admittedly are associated with execution and methodology in studies and evaluations of manufacturing training. Return on investment and the strategic and tactical advantages of training are difficult to factor into formal evaluations, and many companies' perceptions of such training are colored by poor past experience.

Precedent for improving manufacturing skills is lacking in many companies that are unaccustomed to paying for skills beyond entry level. The need to provide for continuing development is slowly being recognized, but managers untrained in modern manufacturing methods are inclined to look at manufacturing training as less productive than other training. Consequently, they remain reluctant to invest in training at the operator and technician levels.

Even with more widespread interest in company-provided training, much of the work force would not be ready to take advantage of it. Problems of basic literacy, necessitating remedial instruction, are impeding the entry into the work force of increasing numbers of educationally and economically disadvantaged people who might otherwise train for the new manufacturing jobs.

A more fundamental barrier to the development of manufacturing skills is the largely indifferent and sometimes negative attitudes of faculty and guidance counselors, most likely derived from lack of manufacturing experience at any level. In some cases, faculty interest in manufacturing is actively discouraged and offending faculty are ostracized. This negative peer pressure can be traced, in part, to a general lack of a scientific base.

Lack of state-of-the-art education facilities constitutes another deficiency in U.S. universities that would teach manufacturing-related science and engineering, according to the DOD report cited earlier. Tax incentives that encouraged donations of industrial equipment have ended, but even with this incentive, manufacturers could not contribute enough equipment to schools to replicate modern manufacturing in all its complexity.

In addition, an appropriate body of manufacturing knowledge in a format suitable for teaching (curricula, content, and study and presentation materials) is lacking at many levels. Because they must compete with established programs for funding, new programs and disciplines focused on manufacturing are extremely difficult to put in place. Given funding (and the effects of the general educational funding malaise are felt by manufacturing as well), such programs will face the task of defining a body of manufacturing knowledge.

The 1988 DOD report finds that the source of the nation's technical skill base, its university system, though sound, has little to offer in manufacturing and manufacturing technology. This is attributed in part to manufacturing's lack of status, even within manufacturing firms, where greater prestige tends to accrue to research and design engineers than to manufacturing engineers. The consequences can be seen in microelectronics manufacturing. To illustrate, the typical undergraduate experience in microelectronics is shown in Table 6-4, and the missing manufacturing engineering content is shown in Table 6-5.[8]

University shortcomings in this field can be grouped into two categories: impressions or attitudes conveyed to students that undermine the country's ability to produce highly talented and skilled semiconductor engineers, and significant curriculum deficiencies that limit students' ability to acquire the broad knowledge needed

TABLE 6-4 Typical Undergraduate Experience in Microelectronics

Course Type	Manufacturing Content
VLSI design (digital systems)	None
Analog integrated circuit design (electronics)	None
Integrated circuit processing lecture	None
Integrated circuit fabrication laboratory	Small
Device physics	None

SOURCE: Microelectronic Engineering at Rochester Institute of Technology: Manpower for Tomorrow's Technology, 1990.

TABLE 6-5 Manufacturing Engineering Content Missing from the Above Curriculum

Subject Area	Specific Skills
Operations Research	Factory floor simulation Work-in-progress tracking Total cycle time management Materials resource planning Scheduling Productive maintenance Joe Juran methodology Gathering and processing data for control and quality improvement*
Statistical Process Control	Design of experiments Statistical thinking Time series analysis Specific training in quality and reliability*
Computer automation	CAD, CAM, CIM SECS I, II Robotics AI, Expert Systems
Other	Lithography

SOURCES: Microelectronic Engineering at Rochester Institute of Technology: Manpower for Tomorrow's Technology, 1990; and Panel on Manufacturing Skills Improvement (*).

to be innovative in this interdisciplinary field. The resulting attitudes and reward systems discourage many of the best candidates from beginning or continuing careers in manufacturing. Finally, the elementary and secondary school teachers who prepare prospective university students are sorely deficient in providing a solid grounding in mathematics and science.

Governmental structure offers one more barrier to improvement of manufacturing skills. Doctrine tends to discourage industry-government involvement in both directions, and federal responsibility for promoting improvement of manufacturing skills, both in industry and for its own use, is unclear. Uncoordinated initiatives are scattered throughout the departments of commerce, defense, education, energy, and labor and the National Aeronautics and Space Administration, National Institute of Standards and Technology, and National Science Foundation (NSF).

RESEARCH NEEDS AND GENERAL RECOMMENDATIONS

Manufacturing skills improvement is particularly difficult because so much of what affects it lies outside the manufacturing sector. The order of the day is to develop competence, in both the work force and management, with the advanced manufacturing technologies essential to manufacturing competitiveness. Very often today in U.S. firms, highly flexible manufacturing systems capable of producing more than one product over the range of economic quantities, and thus of supporting frequent new product introductions and short runs of custom products, are used simply to increase machine utilization for existing small bases of products. Management training in business schools must emphasize technology management at least as much as financial management. Moreover, it must emphasize the development of an integrated view of manufacturing if it is to produce people competent to optimize manufacturing enterprises. Some degree of cross training in engineering would produce managers who have skills comparable to those of their counterparts elsewhere in the world, particularly in Japan.

Also important to the development of a competent work force is vocational and engineering education. At these levels, the problem today is not that skills are not imparted; it is that the *right* skills ar not imparted. What modern manufacturing needs—and is not getting—are master technicians and Renaissance engineers. Identifying the skills needed by these classes of employees in current and future uses of advanced manufacturing technology is an important part of a research agenda. Instruction of such employees should emphasize the application of new ways to improve quality and productivity, such as techniques for robust design, quality programs, production control mechanisms (e.g., Goldratt's and others), and newer accounting systems (e.g., Activity-Based Costing) that derive information from simple, on-floor measurements. As these methods are culture- rather than capital-intensive, engineers must be taught how to introduce them so that they are accepted. Manufacturing engineering programs should give these techniques and their implementation at least as much attention as they give robotics and CAM, which are perhaps more comfortable to deal with technically, but have a more limited range of applicability.

Finally, research should be directed at determining why engineering instruction in the United States tends to emphasize theory over practice and design over production, and why engineering faculty seem to have little manufacturing experience. Determining whether this has always been so or if something happened to turn U.S. university teaching away from a practical production orientation could provide insights useful for restoring the balance.

None of the generic skills noted—such as basic literacy, numeracy, integrative and interpersonal abilities, and problem solving and higher order thinking—are peculiar to manufacturing; these are the skills that virtually every industry, vocational, and skills study has found lacking. Recognition of the indispensability of these skills to people working with advanced manufacturing technologies might serve to reinforce efforts to build them in the population as a whole, but only if manufacturing is itself considered important.

The significance of manufacturing must be understood in terms of what companies do. Companies that manufacture goods must be valued for that activity rather than as chips in some grand national poker game. Manufacturing enterprise must come to be viewed as at least as important as making deals in Wall Street. Until that happens, manufacturing careers will continue to be undervalued and undercompensated.

Many of the efforts that are needed are not clearly within the purview of the NSF. This section attempts to differentiate research activities that, though they might not qualify as research, are activities that the panel believes are crucial to the development and improvement of the manufacturing skills base.

Career-long learning is essential, especially for engineers and technical people in manufacturing. Implementing the priorities below will provide an infrastructure for career-long learning, providing further rationale for funding those priorities.

Education of underrepresented individuals is best implemented by making it an integral part of each of the following priorities.

Priority One: Collaborative Education in Manufacturing Skills

Rationale

A methodology is needed for studying complete industrial processes in order to identify more precisely the skills that will be required by users of advanced technologies. Already, companies have perceived a need for familiarity with statistics, process control, and manufacturing concepts, including microeconomics, basic electronic theory, and communications, as well as problem-solving ability and basic qualities such as responsibility and initiative. A more thorough exposure to these subjects could rely on the establishment of teaching factories, similar in concept to teaching hospitals. Among the many diverse and useful education and research efforts aimed at improving manufacturing skills at the engineering and management end of the education spectrum are: the Laboratory for Manufacturing and Productivity, Materials Processing Center, and Leaders for Manufacturing Program at the Massachusetts Institute of Technology; the manufacturing-oriented Engineering Research Center at Purdue University; the National Technological University's Master of Science Program in Manufacturing Systems Engineering; the Center for Innovation Management Studies and Manufacturing Engineering Program at Lehigh University; and the University of Wisconsin's Manufacturing Engineering initiative. These programs, though they appear to be very effective, are not nearly numerous enough to serve the population in need of skills improvement at this level.

Recommendation

The NSF should establish a program to subsidize the initiation of large new consortia that can collaborate, among themselves and directly with ongoing research efforts, on the development and dissemination of programs of manufacturing skills education for engineers and managers. The effort could be couched as research on collaborative education in manufacturing skills, including nationwide access and hands-on experience at appropriate centers, to include a number of teaching factories.

Comment

Such a program could leverage existing funding from NSF, Defense Advanced Research Projects Agency (DARPA), industry, and universities to mount a larger, more effective effort than exists today. A relatively small amount of funding could have a major impact. Funding could be on a one- or two-year start-up basis, with continuing funding from other sources thereafter.

Priority Two: Educational Level—Faculty

Rationale

Improving the technical competence of students and practicing engineers relies on improving the technical competence of faculty. Engineering faculty should understand the importance of manufacturing and bring manufacturing-related concerns to students at all levels. An effective way to swiftly bring this about is to establish for engineering faculty a professional development program in manufacturing that is cost effective and readily available.

Recommendation

NSF should establish a Faculty Professional Development Program in manufacturing with a goal of reaching 20 percent of the engineering faculty in ABET[9] accredited programs within two years. The proposed program should be operated nationwide, with industry participation, at a relatively low cost per participant per day.

Comment

Existing programs are not effective for this purpose because few faculty from leading institutions attend and the programs are lengthy and costly.

Priority Three: Education of Managers

Rationale

An environment more hospitable to manufacturing will require a change in corporate culture. Manufacturing must become a major concern instead of a secondary consideration of top management. Only management initiative can effect the needed changes in attitude toward and practice of, manufacturing. Studies that show Japanese managements eliciting from the same work force and process technology 15 percent greater productivity and quality than their U.S. counterparts should provide incentive enough. Many U.S. schools of management and business turn out students who understand business; far fewer turn out students who can manage. The concepts of business are learned in a much different way than the behavioral skills needed to manage effectively a sociotechnical system. Alternative sources of the needed training must be found.

Recommendation

NSF should fund and coordinate research that involves business and management schools, engineering colleges, and industry in collaborative studies of manufacturing management in particular and technology management in general. In-house training sponsored by industries and individual companies and educational programs provided through university extension and business and management school executive education programs also are appropriate. As a first step, NSF should fund a program to initiate the collection of data on the current status of programs in manufacturing and technology management.

Priority Four: Establishment of Education and Training Consortia

Rationale

This would be a higher priority if the infrastructure needed to support the program existed at U.S. institutions. Because academe lacks the faculty and facilities to implement needed programs, and the cost of creating a supporting infrastructure would be enormous, NSF should allocate resources to support meaningful cooperation among the major players—institutions, industry, and government.

Recommendation

NSF should work with other government agencies, particularly the Department of Commerce, to develop a program that would establish consortia around specific manufacturing efforts, such as microelectronics, automotive, and aerospace.

Comment

Many elements of such an effort already exist and should be encouraged to expand from a research focus to research and education. The Manufacturing Educational Centers concept is a good direction to follow. The cost is hard to estimate, but could be as high as that of the Engineering Research Centers program (1985/$10 million/6 centers; 1986/$21.89 million/11 centers; 1987/$29.28 million/14 centers; 1988/$33.8 million/16 centers; 1989/$38.17 million/18 centers; 1990/$42.51 million/19 centers; 1991/$45.77 million/19 centers; 1992/$48.27 million/18 centers).

Priority Five: Educational Level—Paraprofessional (Two-year Colleges/Vocational Schools)

Rationale

The funding of research to strengthen the development of paraprofessionals is particularly important to the nation's small and medium-sized businesses, which require broader competencies and greater flexibility and do not have engineering staffs. Much more knowledge is needed about the skill, knowledge, and behavioral requirements of advanced manufacturing systems. Research should be undertaken to help employers and employees articulate needs for specific skills and to help educators translate these needs into curricula. Further, it is clear that the paraprofessional training currently provided by many large companies should be transferred to public institutions. Similarly, firm-specific training being provided by many public agencies ought to be transferred to companies. Support to design and encourage the process for these shifts of responsibility would be appropriate.

Recommendation

To lay the groundwork for future efforts at two-year colleges and vocational schools, successful manufacturing skills and education programs, both domestic and abroad, should be studied for their organizational structures, delivery of services, management, and incentives, as well as for their curricular content. Part of such a study might take the form of an analysis, international in scope, of the organization, funding, execution, and training content of apprenticeship and cooperative programs.

The following priorities are not ranked.

Conceptual Thinking

Rationale

Manufacturing, being easily the most complex and challenging part of an organization, requires high-order conceptual skills. The manufacturing function is most effectively managed as a system, yet most of the thinking brought to bear on manufacturing problems and opportunities is discrete. This is not surprising in light of findings that only a third of students in U.S. colleges and universities have thinking abilities that can be classified as conceptual. The didactic approach to instruction employed in most secondary schools and universities does not prepare students to understand and manage systems. Similarly, the teaching of engineering emphasizes analytical skills, but imparts little in the way of integrative or synthesis skills that will enable students to transcend thinking about discrete parts in order to understand and implement systems.

Recommendation

NSF should fund research to identify, analyze, and document secondary school and university curricula that successfully teach conceptual and integrative thinking. The results of this research should be made widely available, and a panel comprising the educators responsible for designing and delivering the successful programs should be empowered to recommend necessary changes in secondary, university, and graduate curricula.

Cooperative Behavior

Rationale

Although manufacturing is increasingly relying on distributed and cooperative approaches to problem solving, little research has been done on the factors that foster or suppress cooperative behavior. Neither organization of education and training programs nor the philosophy of education have received much attention, though these are at least as important as program content in inculcating work habits. As cooperation and shared responsibility become more important to manufacturing, educators and trainers need to know how their actions and attitudes, as well as the curricula, influence behavior. In addition, research is needed at the human–

machine interface. New approaches to training will be needed to link machine learning capability and operator experience.

Recommendation

Research into the relationships of individuals to complicated systems should be undertaken, to include person–person and person–machine cooperation in both learning and work situations.

Effective and Efficient Teaching Methods

Rationale

Engagement, debate, and student responsibility and participation are teaching methods that shorten the time it takes to learn, as well as lengthen retention, build skills, and facilitate the application of learning.

Recommendation

NSF should fund research on teaching methodologies that yield faster learning and better retention and facilitate the application of what is learned to real life situations. Representatives of both traditional (e.g., universities and community colleges) and nontraditional (e.g., the National Technological University) educational enterprises might be brought together in a financially self-sustaining program to determine how different instructional methods might be combined to bring the highest quality of instruction to the broadest range of students without sacrificing the benefits of individual contact and experimental practice. The unique qualities of television might be studied, for example, with the aim of using it to improve training in manufacturing skills.

Educational Level—Bachelor's Degree

An analysis of manufacturing components in curricula at the bachelor's degree level in engineering and in business education is needed.[10]

Recommendation

NSF should perform an analysis of manufacturing components in curricula at the bachelor's degree level in engineering and in business education.

Educational Level—Through High School

Education through high school is a top national priority that must be addressed with more resources than NSF commands. Some immediate measures are possible. A semiconductor industry recommendation for raising the standards of primary and secondary education suggests that industry experts teach summer sessions as a way of developing mathematics and science teachers. Career education, including the evaluation and development of career guidance materials at all levels, should be undertaken in support of manufacturing. This might be construed to include promoting a general awareness of the importance of manufacturing to the U.S. economy.

Recommendation

NSF should inform those studying the problems of secondary education of its import for manufacturing and should encourage the incorporation of manufacturing awareness in the nation's high school curricula.

Summary

Manufacturing competitiveness in the international arena will rely increasingly on the deployment of advanced manufacturing technology, which in turn will rely on the nurturance of a highly skilled and multidisciplinary work force. This work force does not now exist, and serious impediments, enumerated in this chapter, threaten to slow or even forestall its creation. The consequences of this eventuality for the economic welfare of the United States range from serious to devastating.

Evidence abounds that the U.S. educational system needs radical and pervasive reform. This chapter suggests some strategies for improving the nation's chances of creating the work force it needs. In the interest of starting somewhere, and starting soon, this panel recommends that the following activities be undertaken immediately.

- Establish a program to foster collaborative education in manufacturing skills through large consortia and teaching factories.
- Establish a professional development program to improve faculty understanding of manufacturing and encourage faculty to bring manufacturing concerns to the attention of students.
- Fund collaborative studies of manufacturing and technology man-

agement as a way of developing understanding in managers of the need to change attitudes toward, and the practice of, manufacturing.

- Encourage cooperation among federal, state, and local governments to develop a consortia-based alternative to the educational infrastructure so as to raise graduate education in manufacturing to a satisfactory level.
- Fund an international study of successful programs of training and education in manufacturing skills and fund the transfer of generic teaching functions from industry to academe.

NOTES

1. Gone fishing. 1990. The Economist. 314: (January 6) 61-62.
2. Richards, B. 1990. Wanting Workers. Wall Street Journal Reports: Education (supplement) (February 9) R10.
3. W. B. Johnston and A. H. Packer. 1987. Workforce 2000: Work and workers for the 21st century. Hudson Institute. xxi-xxii, 97-99.
4. Under Secretary of Defense (Acquisition). 1988. Bolstering Defense Industrial Competitiveness. Report to the Secretary of Defense. July. 19-23.
5. Glover, R. W. 1986. Apprenticeship Lessons from Abroad. Columbus, Ohio: The Ohio State University National Center for Research in Vocational Education.
6. A recent National Research Council report recommends establishing such a clearinghouse. See Improving Engineering Design: Designing for Competitive Advantage. 1991. Washington, DC: National Academy Press. 1991.
7. Two exceptions are The Great Lakes Manufacturing Technology Center (supported by the National Institute of Standards and Technology) and the Unified Technology Center in Cleveland, Ohio, which provide focused help for smaller firms. They transfer technology through on-site interactions with smaller firms and enhance employee skills through training. They also provide advice to individual firms on opportunities for enhancing product quality and productivity, which is well received by the business community and is a valuable activity for all concerned.
8. A well-thought-out curriculum for manufacturing and design engineering has been proposed by Dr. Joel Spira, Lutron Electronics Company, Coopersburg, Pennsylvania, January 6, 1990. See Appendix B in Improving Engineering Design: Designing for Competitive Advantage. 1991. Washington, DC: National Academy Press.
9. The Accreditation Board for Engineering and Technology reviews and provides accreditation for the nation's collegiate-level technology-based programs.
10. For DOD activities, see The Department of Defense Report on Science and Engineering Education Activities of the Department of Defense for the Committees on Armed Services, United States Congress. March 1990.

Bibliography

Ad Hoc Steering Committee for the Study of Research Applied to National Needs, Committee on Public Engineering Policy. 1973. Priorities for Research Applicable to National Needs. Washington, D.C.: National Academy of Engineering.

American Society for Engineering Education—Committee on Evaluation of Engineering Education. 1955. Report on Evaluation of Engineering Education. Urbana, Ill.: ASEE. June 15.

Atherton, D. P., and Y. Sunahara, eds. 1990. Control: Theory and Advanced Technology. Tokyo, Japan: Mita Press.

AT&T. 1987. Product Realization. AT&T Technical Journal. 66:(5).

Bailey, T. 1990. Changes in the Nature and Structure of Work: Implications for Skill Requirements and Skill Formation, National Center for Research in Vocational Education. Berkeley, Calif.: University of California, Berkeley.

Bohn, R., and R. Jaikumar. 1989. The Dynamic Approach: An Alternative Paradigm for Operations Management. Harvard Business School Working Paper No. 88011. Boston, Mass.: Revised August 1989.

Boston University Manufacturing Roundtable. 1988. Implementing manufacturing strategies: Breaking the performance measurement barriers. Paper presented at an international conference commemorating the 75th anniversary of the School of Management. October 20-21.

Brand, S. 1987. The Media Lab: Inventing the Future at MIT. New York, N.Y.: Viking Press.

Carnevale, A. P., and J. W. Johnson. 1989. Training America: Strategies for the Nation. Washington, D.C.: American Society for Training and Development.

Chaturvedi, A. R. 1986. Artificial Intelligence in Manufacturing: The State-of-the-Art. Paper submitted to Digital Equipment Corporation,

Hudson, Massachusetts. Milwaukee: University of Wisconsin-Milwaukee.

Clark, K., R. Henderson, and R. Jaikumar. 1989. A Perspective on Computer Integrated Manufacturing Tools. Harvard Business School Working Paper No. 88-048. Boston, Mass. Revised January 1989.

Commission on the Skills of the American Workforce. 1990. America's Choice: High Skills or Low Wages. Rochester, N.Y.: National Commission on Education and the Economy.

Costello, R. B. 1988. Bolstering Defense Industrial Competitiveness: Preserving Our Heritage; The Industrial Base; Securing Our Future. Report to the Secretary of Defense by the Under Secretary of Defense (Acquisition). July 1988.

Cross-Disciplinary Engineering Research Committee and Manufacturing Studies Board—Task Force on Management of Technology. 1987. Management of Technology: The Hidden Competitive Advantage. Washington, D.C.: National Academy Press.

Department of Defense Report on Science and Engineering Education Activities of the Department of Defense for the Committees on Armed Services, United States Congress. March 1990.

Dickson, D. Setting research goals not enough, says OECD. Science 241:898.

Federal Coordinating Council on Science, Engineering and Technology. Annual Report. March 1988.

Feibus, Michael. 1988. Sematech Beckons Brightest: Engineers Going to Promised Land. *San Jose Mercury News*, March 14.

Feigenbaum, E., P. McCorduck, and H. P. Nii. 1988. The Rise of the Expert Company. New York, N.Y.: Times Books.

Flynn, P. 1988. Facilitating Technological Change: The Human Resource Challenge. Cambridge, Mass.: Ballinger Publishing.

Glover, R. W. 1986. Apprenticeship Lessons from Abroad. Columbus, Ohio: The Ohio State University National Center for Research in Vocational Education.

Gone fishing. 1990. *The Economist*. 314: (January 6)

Grygo, G. 1990. Process control on verge of "New Age," report says. Digital Review, January 8, 1990. (Based on recent Yankee Group report.)

Hirschhorn, L. 1984. Beyond Mechanization. Cambridge, Mass.: MIT Press.

Hulthage, I. A. E., M. L. Farinacci, M. S. Fox, and M. D. Rychener. 1990. The Architecture of ALADIN: A Knowledge Based Approach to Alloy Design. *IEEE Expert*. (August).

Imai, M. 1986. Kaizen—The Key to Japanese Competitive Success. New York, N.Y.: Random House.

Jahr, Dale. 1987. Corporate Wealth: More for the Little Guys. *The Wall Street Journal*, Wednesday, January 21.

Jaikumar, R. 1986. Postindustrial Manufacturing. Harvard Business Review (November-December) 69-76. Reprint No. 86606.

Jaikumar, R. 1990. An architecture for a process control costing system. Chapter 7 in Measures for Manufacturing Excellence, R. Kaplan. Boston, Mass.: Harvard Business School Press.

Jaikumar, R., and R. Bohn. 1986. The Development of Intelligent Systems for Industrial Use: A Conceptual Framework. Research on Technological Innovation, Management, and Policy 3:169-211. Boston, Mass.: JAI Press, Inc.

Japan Institute of Plant Maintenance. Establishment of Reliability by TPM. Adeka Argos Chemical Corporation, Mie Factory. (1989 TPM Thesis Award Winning Paper).

Japan Institute of Plant Maintenance. Foreseeing 21st Century Factory Operations Through TPM. Tokai Dainippon Printing Corporation. (1989 TPM Thesis Award Winning Paper).

Japan Institute of Plant Maintenance. Market Adaptability and Competitive Environment Created by TPM. Kubota Steel Corporation, Odawara Factory. (1989 TPM Thesis Award Winning Paper).

Japanese Technology Evaluation Panel on Computer Integrated Manufacturing (CIM) and Computer Assisted Design (CAD) for the Semiconductor Industry in Japan. 1988. Panel report.

Johnston, W. B., and A. H. Packer. 1987. Workforce 2000: Work and workers for the 21st century. Hudson Institute.

Kaplan, R. S. 1989. Management accounting for advanced technological environments. Science 245 (August 25):819-823.

Kaplan, R. S. 1986. Must CIM be justified by faith alone? Harvard Business Review 64(2):87-95.

Kaplan, R. S. 1988. One cost system isn't enough. Harvard Business Review 66(1):61-66.

Koska, D., and Joseph R. 1988. Countdown to the Future: The Manufacturing Engineer in the 21st Century. Profile 21: Executive Summary. Dearborn, Mich.: Society of Manufacturing Engineers.

Levine, M. 1980. High-Definition Television: Time to Take a Stand. Testimony before the Subcommittee on Science, Research and Technology, Hearing on "New Directions in Home Video Technology, June 23.

Lund, R. T., S. R. Rosenthal, and T. Wachtell. 1986. The Manufacturing Executives for the 1990s: Advanced Technology and New Management Skills. Boston, Mass.: Boston University, Center for Technology and Policy.

McKee, K. E., ed. 1987. Automated Inspection and Production Control: Proceedings of 8th International Conference, Springer-Verlag. Bedford, U.K.: IFS Publications, Ltd.

Manufacturing Facilities Reliability Forum, Cleveland Engine Plants. January 1990. Reliability and Maintainability Guidelines: The R&M Approach to Competitiveness. Cleveland, Ohio: The Ford Motor Co.

Manufacturing Technology Information Analysis Center. 1988. Computer Integrated Manufacturing: Concepts and Applications. U.S.

Department of Defense. Washington, D.C. April 1988. Rock Island, Ill.
Miller, J. G., and T. E. Vollmann. 1985. The hidden factory. Harvard Business Review 63(5):142-150.
Murray, A. 1987. The Outlook: The Service Sector's Productivity Problem. *The Wall Street Journal.* February 9.
Nakajima, S. 1988. Introduction to TPM: Total Productive Maintenance. Cambridge, Mass.: Productivity Press Inc.
Nakajima, S. 1988. TPM Development Program: Implementing Total Productive Maintenance. Cambridge, Mass.: Productivity Press Inc.
Nikkan K. Shimbun, Ltd. and Factory Magazine, eds. 1988. Poka-yoke: Improving Product Quality by Preventing Defects. Cambridge, Mass.: Productivity Press Inc.
National Academy of Engineering. 1987. Infrastructure for the 21st Century: Framework for a Research Agenda. Washington, D.C.: National Academy Press.
National Academy of Engineering. 1988. Design and Analysis of Integrated Manufacturing Systems. Washington, D.C.: National Academy Press.
National Academy of Engineering. 1988. Technology in Services: Policies for Growth, Trade, and Employment. Series on Technology and Social Priorities. Washington, D.C.: National Academy Press.
National Academy of Engineering—Committee on Career-Long Education for Engineers. 1988. Focus on the Future: A National Action Plan for Career-Long Education for Engineers. Washington, D.C.: National Academy Press.
National Research Council—Committee on Design Theory and Methodology. 1991. Improving Engineering Design: Designing for Competitive Advantage. Washington, D.C.: National Academy Press.
National Research Council—Committee on the Education and Utilization of the Engineer. 1985. Continuing Education of Engineers. Washington, D.C.: National Academy Press.
National Research Council—Committee on the Education and Utilization of the Engineer. 1985. Engineering Education and Practice in the United States: Foundations of Our Techno-Economic Future. Washington, D.C.: National Academy Press.
National Research Council—Committee on the Education and Utilization of the Engineer. 1985. Engineering Employment Characteristics. Washington, D.C.: National Academy Press.
National Research Council—Committee on the Education and Utilization of the Engineer. 1985. Engineering Graduate Education and Research. Washington, D.C.: National Academy Press.
National Research Council—Committee on the Education and Utilization of the Engineer. 1985. Engineering Infrastructure Diagramming and Modeling. Washington, D.C.: National Academy Press.
National Research Council—Committee on the Education and Utiliza-

tion of the Engineer. 1985. Engineering in Society. Washington, D.C.: National Academy Press.
National Research Council—Committee on the Education and Utilization of the Engineer. 1985. Engineering Technology Education. Washington, D.C.: National Academy Press.
National Research Council—Committee on the Education and Utilization of the Engineer. 1985. Support Organizations for the Engineering Community. Washington, D.C.: National Academy Press.
National Research Council—Committee on the Education and Utilization of the Engineer. 1985. Undergraduate Engineering Education. Washington, D.C.: National Academy Press.
National Research Council—Committee on Science, Engineering, and Public Policy. 1988. Research Briefings 1987. Washington, D.C.: National Academy Press.
National Research Council—Engineering Research Board. 1987. Directions in Engineering Research: An Assessment of Opportunities and Needs. Washington, D.C.: National Academy Press.
National Research Council—Manufacturing Studies Board. 1986. Toward a New Era in U.S. Manufacturing: The Need for a National Vision. Washington, D.C.: National Academy Press.
National Research Council—Manufacturing Studies Board. 1988. The Future of Electronics Assembly: Report of the Panel on Strategic Electronics Manufacturing Technologies. Washington, D.C.: National Academy Press.
National Research Council—Manufacturing Studies Board and Cross-Disciplinary Engineering Research Committee. 1986. Paper from a workshop on "A Framework for Understanding the Ingredients of an Intelligent Knowledge-Based Manufacturing System."
National Research Council—Manufacturing Studies Board and Cross-Disciplinary Engineering Research Committee. 1988. A Research Agenda for CIM: Information Technology. Washington, D.C.: National Academy Press.
National Science Foundation—Division of Science Based Development in Design, Manufacturing and Computer-Integrated Engineering. 1988. Summary of Awards. Fiscal Year 1987.
Office of Technology Assessment. 1983. Automation and the Workplace: A Technical Memorandum. March 1983. Washington, D.C.: U.S. Government Printing Office.
Office of Technology Assessment. 1987. Computerized Manufacturing Automation: Employment, Education, and the Workplace. Washington, D.C.: U.S. Government Printing Office.
Pao, Y.-H. 1989. Adaptive Pattern Recognition and Neural Networks. Reading Mass.: Addison-Wesley.
Pao, Y.-H. 1988. Engineering Artificial Intelligence. Center for Automation and Intelligent Systems Research. Technical Report TR88-107. Cleveland, Ohio: Case Western Reserve University. February 1988.
Pao, Y.-H. 1987. A Perspective on the Evolving Role of Artificial Intelli-

gence Technology in Manufacturing. Center for Automation and Intelligent Systems Research. Technical Report TR87-123. Cleveland, Ohio: Case Western Reserve University. August 1987.

Pao, Y.-H., and D. Sobajic. 1989. Neural Net Technology for Interpretation and Management of Sensor Data. Paper presented at Sensors Expo International, September 13, 1989, Cleveland, Ohio.

Prasad, H. R. Total Productive Maintenance. Forthcoming during 1991.

Research Briefing Panel on Chemical Processing of Materials and Devices for Information Storage and Handling. 1987. Panel report. Washington, D.C.: National Academy Press.

Richards, B. 1990. Wanting Workers. *Wall Street Journal Reports: Education* (supplement) (February 9).

Rosenfeld, S., E. Malizia, and M. Dugan. 1988. Reviving the Rural Factory: Automation and Work in the Rural South. Research Triangle Park, N.C.: Southern Growth Policies Board.

Schmandt, J., and R. Wilson. 1990. Growth Policy in the Age of High Technology. London, England: Unwin Hyman Press.

Schonberger, R. J. 1982. Japanese Manufacturing Techniques. New York, N.Y.: The Free Press.

Senju, S. Concept and Technique of TPM. Japan: Keio University, Engineering Department.

Shingo, S. 1988. Non-Stock Production. Cambridge, Mass.: Productivity Press Inc.

Society for Industrial and Applied Mathematics—Panel on Future Directions in Control Theory. 1988. Future Directions in Control Theory: A Mathematical Perspective. Philadelphia, Pa.: SIAM.

Society of Manufacturing Engineers. 1980. CAD/CAM International Delphi Forecast. Dearborn, Mich.

Society of Manufacturing Engineers. 1982. Directory of Manufacturing Research Needed by Industry. Dearborn, Mich.

Society of Manufacturing Engineers. 1987. Fifth Generation Management for Fifth Generation Technology: A Round Table Discussion. Dearborn, Mich.

Society of Manufacturing Engineers. 1985. Industrial Robots Forecast and Trends: A Second Edition Delphi Study. Society of Manufacturing Engineers, Dearborn, and The University of Michigan, Ann Arbor, Mich.

Taguchi, G. 1988. Introduction to Taguchi methods. Engineering 228:1-2.

Ultratech Stepper. 1988. Research Issues in Deep Ultraviolet. *Lithography*. September 14.

U.S. Air Force Deputy Chief of Staff/Technology and Requirements Planning. 1988. The Air Force Science & Technology and Development Planning Program. June 22.

U.S. Air Force—Machine Tool Task Force. 1980. Technology of Machine Tools. Vols. 1-5.

U.S. Air Force—Office of the Special Assistant for Reliability and Maintainability. 1988. Acquisition Management: USAF R&M 200 Process. Washington, D.C.

U.S. Air Force. 1986. Project Forecast II, The Air Force Tomorrow Team: Executive Summary.

U.S. Air Force Wright Aeronautical Laboratories. 1985. Integrated Information Support System (IISS): An Evolutionary Approach to Integration.

U.S. Congress, House of Representatives. 1989. Creating the New Wealth: National Research and Development Funding in the 1990s. Hearing before the Committee on Science, Space, and Technology, One Hundred First Congress, 1st Session, April 11-12, 1989. Washington, D.C.: U.S. Government Printing Office.

U.S. Congress, Office of Technology Assessment. 1987. International Competition in Services: Banking, Building, Software, Know-how. Washington, D.C.: U.S. Government Printing Office.

U.S. Congress, Office of Technology Assessment. 1987. U.S. Service Industries Threatened by World Competition. Washington, D.C.: U.S. Government Printing Office.

Useem, E. 1986. Low Tech Education in a High Tech World. New York, N.Y.: Free Press.

VerDuin, W. 1990. Neural nets for custom formulation. Proceedings of 19th Annual Programmable Controls/4th Annual Expert System/Industrial Process Control Conference, April 3, 1990, Engineering Society of Detroit. Detroit, Mich.

VerDuin, W. 1990. Neural nets: Software that learns by example. *Computer-Aided Engineering.* (January).

VerDuin, W. 1990. Neural network for diagnosis and control. *Journal of Neural Network Computing.* (Winter).

Wadley, H. N. G., and W. E. Eckart. 1990. Intelligent Processing of Materials, Proceedings of the IMP Symposium held at the 1989 TMS Fall Meeting, Indianapolis, Indiana. Warrendale, Pa.: The Metallurgical Society.

Wadley, H. N. G., P. A. Parris, B. B. Rath, and S. M. Wolf. 1987. Intelligent Processing of Materials and Advanced Sensors. Warrendale, Pa.: The Metallurgical Society.

Warner, M. 1984. Microprocessors, Manpower and Society. New York, N.Y.: St. Martin's Press.

Wireman, T. 1990. World Class Maintenance Management. New York, N.Y.: Industrial Press Inc.

Wood, S., ed. 1988. The Transformation of Work? London, England: Unwin Hyman Press.

Zuboff, S. 1988. In the Age of the Smart Machine: The Future of Work and Power. New York, N.Y.: Basic Books.

APPENDIX

A

Selected Employment Data

TABLE A-1 Civilian Employment in Occupations with 25,000 Workers or More, Actual 1988 and Projected to 2000, Under Low, Medium, and High Scenarios for Economic Growth (Numbers in Thousands)

Occupation	Total employment				1988–2000 employment change					
	1988	2000			Number			Percent		
		Low	Moderate	High	Low	Moderate	High	Low	Moderate	High
Total, all occupations	118,104	127,118	136,211	144,146	9,015	18,107	26,043	8	15	22
Professional specialty occupations	14,628	17,083	18,137	19,072	2,455	3,509	4,444	17	24	30
Engineers	1,411	1,625	1,762	1,933	214	351	522	15	25	37
Aeronautical and astronautical engineers	78	80	88	101	3	10	23	3	13	29
Chemical engineers	49	52	57	62	3	8	13	7	16	27
Civil engineers, including traffic engineers	186	206	219	236	20	32	49	10	17	26
Electrical and electronics engineers	439	565	615	676	126	176	237	29	40	54
Industrial engineers, except safety engineers	132	142	155	171	10	24	40	8	18	30
Mechanical engineers	225	247	269	294	23	44	69	10	20	31
Architects and surveyors	205	227	244	265	22	39	60	11	19	29
Architects, except landscape and marine	86	99	107	117	14	21	31	16	25	36
Surveyors	100	105	112	121	5	12	22	5	12	22
Teachers, librarians, and counselors	5,379	5,937	6,228	6,499	558	849	1,121	10	16	21
Teachers, special education	275	304	317	332	29	43	57	11	16	21
Teachers, preschool	238	290	309	316	53	72	79	22	30	33
Teachers, kindergarten and elementary school	1,359	1,499	1,567	1,638	140	208	279	10	15	21
Teachers, secondary school	1,164	1,328	1,388	1,451	164	224	287	14	19	25
College and university faculty	846	831	869	908	−14	23	63	−2	3	7
Other teachers and instructors	490	514	545	571	24	55	81	5	11	17
Adult and vocational education teachers	467	493	523	548	27	56	81	6	12	17
Instructors, adult (nonvocational) education	227	250	268	282	22	41	54	10	18	24
Teachers and instructors, vocational education and training	239	243	255	266	4	16	27	2	7	11
Librarians, archivists, curators, and related workers	159	168	176	184	9	17	25	6	11	16
Librarians, professional	143	150	157	165	7	14	22	5	10	15
Counselors	124	150	157	164	26	33	41	21	27	33

Engineering and science technicians and technologists	1,273	1,446	1,559	1,690	173	286	417	14	22	33
Engineering technicians	722	858	926	1,007	136	204	285	19	28	39
Electrical and electronic engineering technicians and technologists	341	434	471	515	93	130	174	27	38	51
Drafters	319	331	358	389	12	39	71	4	12	22
Science and mathematics technicians	232	257	275	294	25	43	62	11	19	27
Precision production, craft, and repair occupations	14,159	14,444	15,563	16,683	285	1,404	2,525	2	10	18
Blue-collar worker supervisors	1,797	1,788	1,930	2,074	−9	133	277	−1	7	15
Machinery and related mechanics, installers, and repairers	1,620	1,777	1,910	2,038	157	290	418	10	18	26
Industrial machinery mechanics	463	496	538	580	33	75	117	7	16	25
Maintenance repairers, general utility	1,080	1,199	1,282	1,359	119	202	279	11	19	26
Millwrights	77	83	90	99	6	13	22	8	17	28
Vehicle and mobile equipment mechanics and repairers	1,598	1,738	1,868	1,984	140	270	386	9	17	24
Production occupations, precision	3,190	2,941	3,208	3,453	−249	18	263	−8	1	8
Assemblers, precision	354	236	263	291	−118	−91	−63	−33	−26	−18
Aircraft assemblers, precision	31	28	31	36	−3	−1	5	−11	−2	16
Electrical and electronic equipment assemblers, precision	161	81	91	99	−80	−71	−62	−50	−44	−39
Electromechanical equipment assemblers, precision	59	47	53	58	−11	−6	0	−19	−10	0
Machine builders and other precision machine assemblers	55	42	47	51	−13	−8	−4	−23	−15	−6
Operators, fabricators, and laborers	16,983	15,888	17,198	18,417	−1,095	215	1,434	−6	1	8
Machine setters, set-up operators, operators, and tenders	4,949	4,373	4,779	5,136	−575	−170	187	−12	−3	4
Numerical control machine tool operators and tenders, metal and plastic	64	63	70	77	−1	6	13	−1	9	21
Combination machine tool setters, set-up operators, operators, and tenders	89	88	97	105	−1	8	17	−1	9	19
Machine tool cut and form setters, operators, and tenders, metal and plastic	791	678	747	814	−114	−45	23	−14	−6	3
Drilling and boring machine tool setters and set-up operators, metal and plastic	56	49	54	59	−7	−2	3	−12	−3	6

TABLE A-1 *Continued*

Occupation	Total employment				1988–2000 employment change					
	1988	2000			Number			Percent		
		Low	Moderate	High	Low	Moderate	High	Low	Moderate	High
Grinding machine setters and set-up operators, metal and plastic	72	64	70	77	−8	−1	5	−11	−2	7
Lathe and turning machine tool setters and set-up operators, metal and plastic	89	78	86	94	−11	−3	5	−12	−3	6
Machine forming operators and tenders, metal and plastic	184	151	166	180	−33	−18	−5	−18	−10	−2
Machine tool cutting operators and tenders, metal and plastic	148	121	133	146	−27	−15	−2	−18	−10	−1
Punching machine setters and set-up operators, metal and plastic	51	45	50	54	−6	−1	3	−11	−2	6
Metal fabricating machine setters, operators, and related workers	149	122	134	145	−27	−15	−4	−18	−10	−3
Metal fabricators, structural metal products	40	36	39	42	−4	−1	2	−10	−2	5
Welding machine setters, operators, and tenders	99	78	86	93	−21	−14	−6	−21	−14	−6
Metal and plastic processing machine setters, operators, and related workers	392	363	401	437	−29	9	45	−7	2	11
Electrolytic plating machine operators and tenders, setters and set-up operators, metal and plastic	44	37	41	44	−8	−4	0	−17	−8	0
Metal molding machine operators and tenders, setters and set-up operators	35	31	35	38	−4	−1	2	−12	−2	7
Plastic molding machine operators and tenders, setters and set-up operators	144	159	176	191	15	32	47	11	22	33
Hand workers, including assemblers and fabricators	2,528	2,067	2,266	2,430	−461	−262	−98	−18	−10	−4
Cannery workers	71	63	70	71	−8	−1	−1	−11	−2	−1
Cutters and trimmers, hand	63	59	65	69	−4	2	6	−6	3	10
Electrical and electronic assemblers	237	119	134	144	−118	−103	−93	−50	−44	−39
Grinders and polishers, hand	84	67	74	80	−17	−11	−4	−21	−13	−5
Machine assemblers	47	37	41	45	−9	−5	−2	−20	−12	−4

SOURCE: G. Silvestri and J. Lukasiewicz. 1989. Monthly Labor Review (112:11):51-59.

TABLE A-2 Projected Employment Change by Occupation, 1988-2000, Ranked by Absolute Change in Declining Industries (Numbers in Thousands)

Occupation	Projected 1988–2000 employment change		
	All industries	All declining industries	All growing industries
Total, all occupations	17,120.1	– 1,435.3	18,555.4
All other assemblers and fabricators	– 116.4	– 113.1	– 3.3
Farm workers	– 98.2	– 108.5	10.2
Sewing machine operators, garment	– 90.7	– 96.1	5.4
Inspectors, testers, and graders, precision	– 41.7	– 71.6	29.9
Electrical and electronic assemblers	– 103.3	– 69.0	– 34.3
All other helpers, laborers, and material movers, hand	70.2	– 57.9	128.1
Blue-collar worker supervisors	124.1	– 54.6	178.7
Hand packers and packagers	– 75.0	– 48.8	– 26.2
Secretaries, except legal and medical	383.9	– 44.1	428.0
Electrical and electronic equipment assemblers, precision	– 70.2	– 44.1	– 26.1
Freight, stock, and material movers, hand	19.7	– 37.6	57.3
All other machine operators, tenders, setters, and set-up operators	– 28.5	– 34.1	5.6
Textile draw-out and winding machine operators and tenders	– 30.2	– 30.8	.6
Packaging and filling machine operators and tenders	– 32.6	– 30.1	– 2.5
Child care workers, private household	– 28.1	– 28.1	0
Industrial truck and tractor operators	– 21.4	– 27.6	6.3
Machine feeders and offbearers	– 31.0	– 26.0	– 5.0
Welders and cutters	– 16.1	– 24.8	8.7
Bookkeeping, accounting, and auditing clerks	40.3	– 24.4	64.7
Machine forming operators and tenders, metal and plastic	– 18.4	– 23.4	5.0
General managers and top executives	478.9	– 22.5	501.4
All other hand workers	– 18.5	– 19.6	1.1
All other mechanics, installers, and repairers	– 25.9	– 17.7	– 8.3
Gardeners and groundskeepers, except farm	149.4	– 17.5	166.9
Janitors and cleaners, including maids and housekeeping cleaners	471.8	– 16.8	488.6
Crushing and mixing machine operators and tenders	– 18.9	– 15.8	– 3.1
Sewing machine operators, nongarment	– 8.0	– 15.3	7.2
Machine tool cutting operators and tenders, metal and plastic	– 14.9	– 14.1	– .8
Typists and word processors	– 66.2	– 13.3	– 52.9
Welding machine setters, operators, and tenders	– 13.6	– 13.0	– .6
Cleaners and servants, private household	– 12.6	– 12.6	0
All other metal and plastic machine setters, operators, and related workers	– 11.5	– 11.9	.3
General office clerks	454.3	– 11.1	465.4
All other machine tool cutting and forming, etc.	– 4.3	– 10.9	6.5
Chemical equipment controllers, operators, and tenders	– 10.8	– 10.2	– .6
Sheet metal workers and duct installers	9.7	– 10.1	19.8

SOURCE: G. Silvestri and J. Lukasiewicz. 1989. Monthly Labor Review (112:11):61.

TABLE A-3 Percent Change in Employment for Selected Occupations, 1988-2000, and Percent of Employment Comprised by Whites, Blacks, and Hispanics, 1988

Occupation	Percent change, 1988–2000	Percent comprised by—		
		Whites	Blacks	Hispanics
Total, all occupations	15	87	10	7
Executive, administrative, and managerial occupations	22	92	6	4
Professional specialty occupations	24	89	7	3
Engineers	25	90	4	3
Computer, mathematical, and operations research analysts	52	86	7	3
Natural scientists	19	90	3	3
Health diagnosing occupations	24	88	3	4
Health assessment occupations	38	87	8	3
Teachers, college	3	89	4	4
Teachers, except college	18	89	9	4
Lawyers and judges	30	96	2	2
Other professional workers	23	90	8	4
Technicians and related support occupations	32	86	9	4
Health technicians and technologists	34	81	14	4
Engineering and scientific technicians	22	89	7	5
All other technicians	39	88	7	4
Marketing and sales occupations	20	91	6	5
Administrative support occupations, including clerical	12	86	11	6
Clerical supervisors and managers	12	85	14	6
Computer operators and peripheral equipment operators	29	83	14	6
Secretaries, typists, and stenographers	10	89	8	5
Financial recordkeeping occupations	1	90	6	5
Mail clerks and messengers	10	74	22	9
Other clerical occupations	13	84	13	7
Service occupations	23	79	18	10
Private household workers	−5	76	23	17
Protective service occupations	23	81	17	6
Food service occupations	23	83	12	10
Health service occupations	34	69	28	6
Cleaning service occupations	20	74	23	15
Personal service occupations	27	85	12	8
Precision production, craft, and repair occupations	10	90	8	8
Mechanics, installers, and repairers	13	91	7	8
Construction trades	16	91	7	8
Other precision production occupations	3	88	8	9
Operatives, fabricators, and laborers	1	82	15	11
Machine setters, set-up operators, operators, and tenders	−3	83	15	7
Transportation and material moving machine and vehicle operators	12	82	16	11
Helpers, laborers, and material movers, hand	2	82	15	13
Agriculture, forestry, fishing, and related workers	−5	92	7	13

NOTE: Hispanics can be of any race.

SOURCE: G. Silvestri and J. Lukasiewicz. 1989. Monthly Labor Review (112:11):64.

TABLE A-4 Percent Distribution of Employment by Occupation, 1988 and Projected 2000 Alternatives

Occupation	1988	2000		
		Low	Moderate	High
Total employment	100.0	100.0	100.0	100.0
Executive, administrative, and managerial occupations	10.2	10.8	10.8	10.9
Professional specialty occupations	12.4	13.4	13.3	13.2
Techicians and related support occupations	3.3	3.7	3.7	3.7
Marketing and sales occupations	11.3	11.6	11.7	11.6
Administrative support occupations, including clerical	17.8	17.4	17.3	17.3
Service occupations	15.6	16.7	16.6	16.4
Agriculture, forestry, fishing, and related occupations	3.0	2.4	2.4	2.5
Precision production, craft, and repair occupations	12.0	11.4	11.4	11.6
Operators, fabricators, and laborers	14.4	12.5	12.6	12.8

SOURCE: G. Silvestri and J. Lukasiewicz. 1989. Monthly Labor Review (112:11):65.

APPENDIX

B

Panels of the Committee on Analysis of Research Directions and Needs in U.S. Manufacturing

PANEL ON INTELLIGENT MANUFACTURING CONTROL

AVAK AVAKIAN, *Cochairman*, Vice President (retired), GTE Government Systems, Concord, Massachusetts

RAMCHANDRAN JAIKUMAR, *Cochairman*, Professor of Business Administration, Graduate School of Business Administration, Harvard University, Boston, Massachusetts

CHRISTOPHER S. FUSELIER, Manager, Cell Control Programs, GE Fanuc Automation North America, Inc., Charlottesville, Virginia

YU-CHI HO, Gordon McKay Professor of Engineering and Applied Mathematics, Harvard University, Cambridge, Massachusetts

DAVID A. HODGES, Professor of Electrical Engineering and Computer Sciences, University of California, Berkeley

JOEL MOSES, Dean of Engineering, Massachusetts Institute of Technology, Cambridge

DAVID W. RUDY, Information Systems Liaison to Manufacturing, E.I. du Pont de Nemours & Company, Wilmington, Delaware

GEORGE N. SARIDIS, Professor, Electrical Computer Systems Engineering Department, Rensselaer Polytechnic Institute, Troy, New York

ED SHAMAH, Manager of Advanced Manufacturing Systems, Intel Corporation, Chandler, Arizona

HARRY E. STEPHANOU, Director, Center for Advanced Technology in Automation and Robotics, Rensselaer Polytechnic Institute, Troy, New York

Liaisons to the National Center for Manufacturing Sciences

JONATHAN GOLOVIN, Chairman, Consilium, Inc., Mountain View, California

THOMAS L. HAYNES, Senior Member Technical Staff, Industrial Systems Division, Texas Instruments, Inc., Johnson City, Tennessee

PANEL ON EQUIPMENT RELIABILITY AND MAINTENANCE

STEVEN J. BOMBA, *Cochairman*, Vice President of Technology, Johnson Controls, Inc., Milwaukee, Wisconsin
HRIDAY R. PRASAD, *Cochairman*, Manager of Technology Planning, North American Automotive Manufacturing Operations, Ford Motor Company, Dearborn, Michigan
GIFFORD M. BROWN, Plant Manager, Engine Division, Ford Motor Company, Brook Park, Ohio
ROBERT E. BRUCK, Manager, Corporate Capital Acquisition, Intel Corporation, Chandler, Arizona
CHARLES E. EBERLE, Executive Vice President, Consumer Products Business, James River Corporation, Richmond, Virginia
JOSEPH T. KAMMAN, Director of Technical Services, Cincinnati Milacron, Inc., Cincinnati, Ohio
P. RANGANATH NAYAK, Vice President, Operations Management, Arthur D. Little, Inc., Cambridge, Massachusetts
MICHAEL H. O'NEAL, Director, Operations Research, Northern Telecom, Inc., Nashville, Tennessee
HENRY W. STOLL, Technical Director—Design Technology, Square D Company, Palatine, Illinois
KHALIL S. TARAMAN, Dean, School of Engineering, Lawrence Technological University, Southfield, Michigan
ROBERT L. YEATON, Manager, Manufacturing Automation Technology, GE Aircraft Engines, Lynn, Massachusetts

Liaisons to the National Center for Manufacturing Sciences

LEONARD ALLGAIER, Manager, Power Train Manufacturing, GM Technology Center, Warren, Michigan
GRAHAM D. COTTAM, Vice President, H. R. Krueger Machine Tool, Inc., Farmington, Michigan

PANEL ON MANUFACTURING OF AND WITH ADVANCED ENGINEERED MATERIALS

MELVIN BERNSTEIN, *Cochairman*, Provost and Academic Vice President, Illinois Institute of Technology, Chicago
R. BYRON PIPES, *Cochairman*, Dean, College of Engineering, University of Delaware, Newark
DANIEL G. BACKMAN, Manager, Intelligent Processing of Materials Technology, GE Aircraft Engines, Lynn, Massachusetts
H. KENT BOWEN, Ford Professor of Engineering, Massachusetts Institute of Technology, Cambridge
JOEL P. CLARK, Professor, Department of Materials Science and Engineering, Massachusetts Institute of Technology, Cambridge
THOMAS C. McGILL, Fletcher Jones Professor of Applied Physics, California Institute of Technology, Pasadena
MICHAEL V. NEVITT, Senior Metallurgist, Materials Science Division, Argonne National Laboratory, Argonne, Illinois
DAVID J. SROLOVITZ, Associate Professor, Materials Science and Engineering and Applied Physics, University of Michigan, Ann Arbor
BEN G. STREETMAN, Director, Microelectronics Research Center, University of Texas, Austin
KURT F. WISSBRUN, Senior Research Associate, Hoechst-Celanese Corporation, Summit, New Jersey

Liaison to the Department of Defense

VINCENT J. RUSSO, Director, Air Force Wright Aeronautical Laboratory, Wright-Patterson Air Force Base, Ohio

Liaison to the Committee

JAMES C. WILLIAMS, Department Manager, Engineering Materials Technology Laboratory, General Electric Company, Cincinnati, Ohio

Liaisons to the National Center for Manufacturing Sciences

CHARLES BEETZ, Director, Thin Films R&D, Advanced Technology Materials, Inc., New Milford, Connecticut
FOSTER LAMM, Senior Research Scientist, United Technologies Research Laboratories, East Hartford, Connecticut

PANEL ON RAPID PRODUCT REALIZATION PROCESS

A. TIM SHERROD, *Cochairman*, President, Savant Solutions Company, Menlo Park, California

MICHAEL J. WOZNY, *Cochairman*, Director, Rensselaer Design Research Center, Rensselaer Polytechnic Institute, Troy, New York

HARRIS M. BURTE, Chief Scientist, Air Force Materials Laboratory, Wright-Patterson Air Force Base, Ohio

DON P. CLAUSING, Bernard M. Gordon Adjunct Professor of Engineering Innovation and Practice, Massachusetts Institute of Technology, Cambridge

GEORGE FOO, Director, Manufacturing and Engineering Director, AT&T Company, Little Rock, Arkansas

MARK S. FOX, Associate Professor of Computer Science and Robotics, Carnegie Mellon University, Pittsburgh, Pennsylvania

GARY MARKOVITS, Program Manager for Patent Processes, IBM Corporation, Stormytown, Yorktown Heights, New York

STUART G. MILLER, Manager, Automation Systems Laboratory, General Electric Company, Schenectady, New York

JAMES P. PRENDERGAST, Engineering Manager, Intel Corporation, Chandler, Arizona

FRIEDRICH B. PRINZ, Professor, Department of Mechanical Engineering, Carnegie Mellon University, Pittsburgh, Pennsylvania

JERRY L. PYLES, Manager, Architecture and Standards, SEMATECH, Austin, Texas

Liaison to the Committee

GUSTAV J. OLLING, Chief, Automotive Research and CAD/CAM User Systems, Chrysler Corporation, Highland Park, Michigan

Liaisons to the National Center for Manufacturing Sciences

DANIEL J. MAAS, Manager, State of the Art Assessments, National Center for Manufacturing Sciences, Ann Arbor, Michigan

LARRY McARTHUR, President and Chief Executive Officer, Aries Technology, Inc., Lowell, Massachusetts

PANEL ON MANUFACTURING SKILLS IMPROVEMENT

GERARDO BENI, *Cochairman*, Director, Center for Robotics Systems in Microelectronics, University of California, Santa Barbara
WILLIAM G. HOWARD, JR., *Cochairman*, Senior Fellow, National Academy of Engineering, Scottsdale, Arizona
CARLTON BRAUN, Vice President and Director, Motorola Management Institute, Schaumburg, Illinois
R. SCOTT FOSLER, Vice President and Director of Government Studies, Committee for Economic Development, Washington, D.C.
PHILIP H. FRANCIS, Vice President, Corporate Technical Center, Square D Company, Palatine, Illinois
C. ROLAND HADEN, Dean, College of Engineering and Applied Sciences, Arizona State University, Tempe
JACQUES KOPPEL, President, Greater Minnesota Corporation, Minneapolis, Minnesota
ROY H. MATTSON, Academic Vice President, National Technical University, Fort Collins, Colorado
STUART A. ROSENFELD, Deputy Director, The Southern Growth Policies Board, Research Triangle Park, North Carolina
JAMES J. SOLBERG, Engineering Research Center for Intelligent Manufacturing Systems, Purdue University, West Lafayette, Indiana
F. KARL WILLENBROCK, Assistant Director, Scientific, Technological, and International Affairs, National Science Foundation, Washington, D.C.

Liaisons to the National Center for Manufacturing Sciences

PETER D. HALL, Director of Asia-Pacific Sales, The Gleason Works, Rochester, New York
ALFRED D. ZEISLER, Manager, Manufacturing Systems, AT&T Company, Berkeley Heights, New Jersey

PANEL ON ALTERNATIVE CONCEPTS IN MANUFACTURING

DAN L. SHUNK, *Chairman*, Director, CIM Systems Research Center, Arizona State University, Tempe

MICHAEL G. BORRUS, Deputy Director, Berkeley Roundtable on the International Economy, University of California, Berkeley

CLINTON W. KELLY III, Corporate Vice President, Advanced Technology Programs, Science Applications International Corporation, McLean, Virginia

MICHAEL M. KUTCHER, Manufacturing Consultant, IBM Corporation, Kingston, New York

THOMAS J. LINDEM, Vice President of Technology, The Ingersoll Milling Machine Company, Rockford, Illinois

CHARLES R. McLEAN, Computer Scientist, Group Leader, Production Management Systems, Factory Automated Systems Division, National Institute of Standards and Technology, Gaithersburg, Maryland

GARY L. MICHAELSON, Vice President of Operations, Boeing Commercial Airplane Group, Wichita Division, Wichita, Kansas

ROGER N. NAGEL, Harvey Wagner Professor of Manufacturing Systems Engineering, Lehigh University, Bethlehem, Pennsylvania

PHILIP A. PARRISH, Vice President, Concurrent Engineering Technology, BDM International, Inc., Arlington, Virginia

Liaison to the Industrial Technology Institute

GEORGE H. KUPER, President, Industrial Technology Institute, Ann Arbor, Michigan

Liaisons to the National Center for Manufacturing Sciences

GARY W. COTTEN, Production Engineering Manager, Texas Instruments, Inc., Lewisville, Texas

E. DENNIS WISNOSKY, President, Wizdom Systems, Inc., Naperville, Illinois

Index

A

F

G

H

I

M

X

Z